AF758142

ELSEVIER OCEAN ENGINEERING BOOK SERIES

VOLUME 8

HURRICANE-GENERATED SEAS

Elsevier Internet Homepage: http://www.elsevier.com

Consult the Elsevier homepage for full catalogue information on all books, journals and electronic products and services.

Ocean Engineering Series

Published

WATSON
Practical Ship Design
ISBN: 008-042999-8

YOUNG
Wind Generated Ocean Waves
ISBN: 008-043317-0

BAI
Pipelines and Risers
ISBN: 008-043712-5

JENSEN
Load and Global Response of Ships
ISBN: 008-043953-5

TUCKER & PITT
Waves in Ocean Engineering
ISBN: 008-043566-1

BAI
Marine Structural Design
ISBN: 008-043921-7

PILLAY & WANG
Technology and Safety of Marine Systems
ISBN: 008-044148-3

Forthcoming titles

BOSE & BROOKE
Wave Energy Conversion
ISBN: 008-044212-9

KOBYLINSKI & KASTNER/
BELENKY & SEVASTIANOV
Stability and Safety of Ships (in two volumes)
ISBN: 008-044355-9 (Set)
ISBN: 008-043001-5 (Volume 9)
ISBN: 008-044354-0 (Volume 10)

MUNCHMEYER
Boat and Ship Geometry
ISBN: 008-043998-5

McCORMICK & KRAEMER
Ocean Wave Energy Utilization
ISBN: 008-043932-2

Related Journals
Free specimen copy gladly sent on request. Elsevier Ltd, The Boulevard, Langford Lane, Kidlington, Oxford, OX5 1GB, UK

Applied Ocean Research
Advances in Engineering Software
CAD
Coastal Engineering
Composite Structures
Computers and Structures
Construction and Building Materials
Engineering Failure Analysis
Engineering Fracture Mechanics
Engineering Structures
Finite Elements in Analysis and Design
International Journal of Solids and Structures
Journal of Constructional Steel Research
Marine Structures
NDT & E International
Ocean Engineering
Structural Safety
Thin-Walled Structures

To Contact the Publisher
Elsevier welcomes enquiries concerning publishing proposals: books, journal special issues, conference proceedings, etc. All formats and media can be considered. Should you have a publishing proposal you wish to discuss, please contact, without obligation, the publisher responsible for Elsevier's civil and structural engineering publishing programme:

James Sullivan
Publishing Editor
Elsevier Ltd
The Boulevard, Langford Lane
Kidlington, Oxford
OX5 1GB, UK

Phone: +44 1865 843178
Fax: +44 1865 843920
E-mail: j.sullivan@elsevier.com

General enquiries, including placing orders, should be directed to Elsevier's Regional Sales Offices – please access the Elsevier homepage for full contact details (homepage details at the top of this page).

ELSEVIER OCEAN ENGINEERING BOOK SERIES

VOLUME 8

HURRICANE-GENERATED SEAS

MICHEL K. OCHI

University of Florida
Gainesville, Florida, USA

OCEAN ENGINEERING SERIES EDITORS

R. Bhattacharyya

US Naval Academy
Annapolis, MD, USA

M.E. McCormick

US Naval Academy
Annapolis, MD, USA

ELSEVIER

2003

Amsterdam – Boston – Heidelberg – London – New York – Oxford – Paris
San Diego – San Francisco – Singapore – Sydney – Tokyo

ELSEVIER Ltd
The Boulevard, Langford Lane
Kidlington, Oxford OX5 1GB, UK

© 2003 Elsevier Ltd. All rights reserved.

This work is protected under copyright by Elsevier, and the following terms and conditions apply to its use:

Photocopying
Single photocopies of single chapters may be made for personal use as allowed by national copyright laws. Permission of the Publisher and payment of a fee is required for all other photocopying, including multiple or systematic copying, copying for advertising or promotional purposes, resale, and all forms of document delivery. Special rates are available for educational institutions that wish to make photocopies for non-profit educational classroom use.

Permissions may be sought directly from Elsevier's Science & Technology Rights Department in Oxford, UK: phone: (+44) 1865 843830, fax: (+44) 1865 853333, e-mail: permissions@elsevier.com. You may also complete your request on-line via the Elsevier homepage (http://www.elsevier.com), by selecting 'Customer Support' and then 'Obtaining Permissions'.

In the USA, users may clear permissions and make payments through the Copyright Clearance Center, Inc., 222 Rosewood Drive, Danvers, MA 01923, USA; phone: (+1) (978) 7508400, fax: (+1) (978) 7504744, and in the UK through the Copyright Licensing Agency Rapid Clearance Service (CLARCS), 90 Tottenham Court Road, London W1P 0LP, UK; phone: (+44) 207 631 5555; fax: (+44) 207 631 5500. Other countries may have a local reprographic rights agency for payments.

Derivative Works
Tables of contents may be reproduced for internal circulation, but permission of Elsevier is required for external resale or distribution of such material.
Permission of the Publisher is required for all other derivative works, including compilations and translations.

Electronic Storage or Usage
Permission of the Publisher is required to store or use electronically any material contained in this work, including any chapter or part of a chapter.

Except as outlined above, no part of this work may be reproduced, stored in a retrieval system or transmitted in any form or by any means, electronic, mechanical, photocopying, recording or otherwise, without prior written permission of the Publisher.
Address permissions requests to: Elsevier's Science & Technology Rights Department, at the phone, fax and e-mail addresses noted above.

Notice
No responsibility is assumed by the Publisher for any injury and/or damage to persons or property as a matter of products liability, negligence or otherwise, or from any use or operation of any methods, products, instructions or ideas contained in the material herein. Because of rapid advances in the medical sciences, in particular, independent verification of diagnoses and drug dosages should be made.

First edition 2003

Library of Congress Cataloging in Publication Data
A catalog record from the Library of Congress has been applied for.

British Library Cataloguing in Publication Data
A catalogue record from the British Library has been applied for.

ISBN: 0 08 044312 5

♾ The paper used in this publication meets the requirements of ANSI/NISO Z39.48-1992 (Permanence of Paper).
Printed in The Netherlands.

SERIES PREFACE

In this day and age, humankind has come to the realization that the Earth's resources are limited. In the 19th and 20th Centuries, these resources have been exploited to such an extent that their availability to future generations is now in question. In an attempt to reverse this march towards self-destruction, we have turned our attention to the oceans, realizing that these bodies of water are both sources for potable water, food and minerals and are relied upon for World commerce. In order to help engineers more knowledgeably and constructively exploit the oceans, the **Elsevier Ocean Engineering Book Series** has been created.

The **Elsevier Ocean Engineering Book Series** gives experts in various areas of ocean technology the opportunity to relate to others their knowledge and expertise. In a continual process, we are assembling world-class technologists who have both the desire and the ability to write books. These individuals select the subjects for their books based on their educational backgrounds and professional experiences.

The series differs from other ocean engineering book series in that the books are directed more towards technology than science, with a few exceptions. Those exceptions we judge to have immediate applications to many of the ocean technology fields. Our goal is to cover the broad areas of naval architecture, coastal engineering, ocean engineering acoustics, marine systems engineering, applied oceanography, ocean energy conversion, design of offshore structures, reliability of ocean structures and systems and many others. The books are written so that readers entering the topic fields can acquire a working level of expertise from their readings.

We hope that the books in the series are well-received by the ocean engineering community.

Rameswar Bhattacharyya
Michael E. McCormick

Series Editors

Preface

Hurricanes (typhoons in the UK and Asia) are one of the most adverse aspects of the ocean environment which may give rise to a disastrous event for marine systems in the ocean. Furthermore, when a hurricane advances toward the shoreline it causes immense destruction on habitation and the natural environment in the vicinity of the landing site. A considerable amount of information on hurricane winds has been presented to date. Nevertheless, only limited information is currently available on hurricane-generated seas, although many studies on severe seas generated by ordinary winds have been carried out.

It is of interest to note that the characteristics of hurricane-generated seas in the open ocean are quite different from those of the usual wind-generated seas in that the source of potential energy generating the waves is advancing in the same direction as the propagating waves. As a result, the wave spectrum representing the sea severity at a site located in the path of an advancing hurricane sharply increases in magnitude (particularly, the low frequency components) 3 to 4 hours prior to the wind speed becoming maximum at the site. Such a rapid and intense change in the wave spectrum, not observed in ordinary wind-generated storms, results in the wave spectrum of hurricane-generated seas being unique in shape. Hence, clarification of the severity and characteristics of wave spectra associated with hurricane winds in the open ocean may provide information vital for avoiding disastrous events experienced by ships and offshore structures.

At the time of hurricane landing, intense sea severity near the shoreline results in the beach and dune system being underwater, and causes a devastating impact to the residents, structures and natural environment. Quantitative evaluation of the extreme nearshore sea condition at the time of hurricane landing for a specified wind speed would contribute significantly for the improvement of technological practice and management decision making regarding the conservation of coastal and marine resources including dune systems and buildings near the shoreline.

With the prospective goal stated in the preceding, this book is designed as a reference for researchers, designers, graduate students and technical managers in naval, ocean and coastal engineering. It presents information on the severity of hurricane-generated seas in the deep ocean, in finite water depth of the continental shelf as well as nearshore at the time of hurricane landing. Most wave data of hurricane-generated seas in the deep ocean referred to in this text are measured by NOAA buoys while some data are obtained by aircraft equipped with measurement devices presented in the researcher's references. On the other hand, information on the sea severity in finite water depth and nearshore is estimated by employing methodology developed based on the stochastic process approach. The validity of the methodology is confirmed through comparison of the estimated results with available measured data obtained during severe storms.

This book consists of five chapters. Chapter 1 addresses hurricane winds and the resultant generated sea states in the open ocean. The difference between hurricane wind spectra over the ocean and those over land is clarified. Turbulent wind spectra obtained from measured data over the sea are compiled and the average spectrum is presented in a mathematical formulation. Attention is given to the significance of turbulent wind energy for design of marine systems having low frequency response characteristics in a seaway such as surging motion of the tension-leg platform. The relationship between hurricane wind speed and the generated sea severity (significant wave height) is presented from analysis of data obtained by NOAA buoys. Emphasis is given to the fact that a hurricane-generated sea depends not only on the magnitude of wind speed but also, to a great extent, on the sea condition during the 7–10 days prior to the hurricane arrival at the site.

Chapter 2 discusses hurricane-generated seas in deep water from the stochastic view point based on the results of analysis of measured data. Unique features of wave spectra observed during hurricane-generated seas are presented and compared with wave spectra observed in ordinary wind-generated seas. A mathematical formulation specifically applicable to hurricane-generated seas is presented in the form of the JONSWAP spectral formulation. The parameters involved in the formulation are determined based on the results of analysis of many spectra of measured data obtained during hurricanes and presented as a function of significant wave height and modal frequency. The procedure for estimating the wave spectrum at a specific site located in the path of an approaching hurricane from knowledge of the atmospheric pressure is also discussed. Additionally, the directional characteristics of hurricane-generated seas obtained by aircraft equipped with NOAA (NASA) airborne synthetic aperture radar (SAR) or scanning radar altimeter (SRA) systems during hurricanes are discussed.

Chapter 3 addresses the transformation of wave spectra with the advance of a hurricane from deep to shallow water. The transformation is presented in three phases. In the first phase, a method for estimating the transformation of wave spectra developed by Kitaigorodskii is discussed with application to the hurricane advancing from deep sea to finite water depth over the continental shelf. A method to estimate the water at which the effect of depth initiates is discussed with an example of computations applied to hurricane KATE.. Estimation of wave spectra on the continental shelf so obtained is valid until the waves reach a certain water depth where wave breaking takes place, called the breaking point. The second phase discusses in detail a method for evaluating the location of the breaking point in random seas, taking the probability of occurrence of breaking as well as the reduction of energy in the wave spectrum into consideration. Finally, estimation of wave spectra in the surf zone defined here as the region extending from the breaking point to the shoreline is discussed.

Chapter 4 discusses wave characteristics including extreme height expected during passage of a hurricane from deep to shallow water. A very important and interesting subject pertains to wave characteristics in finite water depth in which nonlinearity is introduced and thereby the stochastic properties of waves are transformed from Gaussian to a non-Gaussian random process. First, the limiting water depth for a given sea severity below which waves are considered to be a non-Gaussian random process is clarified. The results of computations made for hurricane KATE indicate that the effect of finite water depth initiates at 120 m depth on the continental shelf, but the limiting water depth below which waves are considered to be non-Gaussian is 33.7m. Second, the probability distribution of the wave profile (displacement from the mean value), peak and trough amplitudes as well as wave height applicable to non-Gaussian waves are presented with numerical examples as applied to hurricane KATE.

Chapter 5 addresses nearshore sea severity at the time of hurricane landing. As an example, the probability distribution of wave height in hurricane KATE is obtained from the wave spectrum at each water depth estimated from knowledge of the measured wave spectrum in deep water. The significance of the effect of increase in sea level due to storm surge and/or tide on the nearshore sea severity is discussed with a numerical example. The severity of the nearshore sea condition when a hurricane approaches a specific site depends largely on the bottom profile (slope) of the site; the milder the slope, the less severe the sea condition. In order to represent the sea severity near the shoreline at the time of hurricane landing a dimensionless indicator, called the landing sea severity indicator, is defined. The indicator may be used as a criterion to evaluate the relative severity of sea state assuming a hurricane having the same strength lands at different sites. At the time of hurricane landing, the beach and part of the dune system will be underwater due to storm surge and/or tide. A method to estimate the onshore sea severity is discussed by employing a simple mathematical model representing the beach and dune system. As an example, the results of computations made for hurricane KATE landing at a specified site is presented.

The appendix presents a brief explanation of the derivation of the probability density functions of displacement, amplitudes (peak and trough) and height of non-Gaussian waves necessary for estimating wave characteristics in finite water depth. The probability density functions consist of three parameters. Methods for evaluating these three parameters from either the wave record or the wave spectrum are explained.

Acknowledgements

I am grateful to the College of Engineering, University of Florida, for granting me sabbatical leave to write this book. The majority of the work on hurricane-generated seas was carried out under the sponsorship of the Florida Sea Grant College Program. The generous support of the program committee; in particular Dr. Seaman for his interest and encouragement is greatly appreciated.

Heartfelt thanks go to Professor McCormick and Professor Bhattacharyya, friends for many years, for their advice and encouragement in preparation of this book.

I gratefully acknowledge my graduate students who worked on projects associated with hurricane-generated seas for their patience and diligent effort in analyzing the enormous amount of data required for the culmination of this work. They are, Jayesh Antani, Ming-Hsing Chiu, David Wang, Emmett Foster, David Robillard, William Finlayson and Robert McClellan. I extend sincere thanks to Mr. Malakar for his diligence in the development of computer programs for analysis of a wide variety of data. Finally, it is my pleasant duty to express sincere gratitude to my wife Margaret who read through the entire manuscript and provided valuable assistance in editorial work.

List of Tables

Contents

1

Hurricane wind and sea state in the ocean

1.1 Hurricane wind over the ocean

1.1.1 Introduction

It is commonly known that winds (not necessarily associated with severe storms) are more severe over the ocean than on land because of the wide open space. As will be shown later, the wind spectrum representing the wind energy as a function of frequency clearly shows that the magnitude of energy density at low frequencies is much larger for winds over the ocean than over land. To elaborate on the low frequency wind components, Figure 1.1 taken from Shiraishi's publication (Shiraishi, 1960) shows an example of the time history of wind speed. As seen in the figure, wind speed consists of a variety of frequencies. High frequency turbulence, often called gust winds, give an impression that the energy is large but in reality their energy is relatively small. On the other hand, low frequency components such as those indicated by the dotted lines in the figure representing two different frequencies carry substantially large amounts of energy; in particular, the lowest of these two frequency components does so.

It is noted that the natural response frequencies of some marine systems such as those of surging or yawing motions of a tension-leg platform or a moored ship appear to be in the frequency domain where a very large wind energy is observed. When this is the case, winds (particularly hurricane winds over the seaway) may induce resonance in addition to the severe response produced by rough waves, resulting in a critical condition for the system. Hence, for the safety of marine systems operating in a seaway, emphasis is placed on the low frequency components of wind energy.

Prior to discussing wind characteristics in detail, it may be helpful to briefly address the formation of hurricane winds. Hurricanes (called typhoons in the UK and Asia) are defined as tropical cyclones with average sustained wind speeds exceeding 75 mph (65 knots, 33 m/s).

Every hurricane causes disastrous nearshore and onshore damage at the time of its landing. Consequently, a damage potential scale consisting of five categories indicating

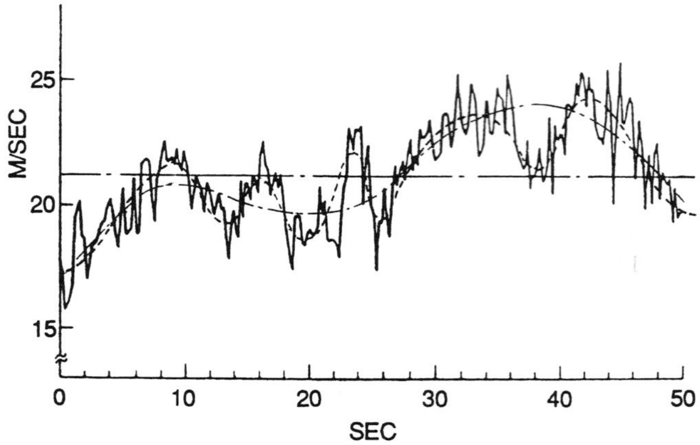

Figure 1.1 – Example of time history of wind speed (Shiraishi, 1960).

the severity of the hurricane is often used for public information. This scale is called the "Saffir–Simpson scale" given in Table 1.1.

Hurricanes (tropical cyclones) originate over tropical oceans in latitudes between 5 and 20° from the equator where the surface water temperature is sufficiently warm, 77 °F (27 °C), or warmer. Hurricanes (tropical cyclones) can form on either side of but not within 5° of the equator. A strong low pressure pumps up a large amount of air from just above the sea surface along with an enormous amount of heat energy into the atmosphere. Warm moist air rapidly rises in spiral fashion and condenses to form clouds and precipitation in the upper atmosphere. As the air flows inward, it is deflected by the Coriolis effect which flows counter-clock wise in the Northern hemisphere (clock wise in the Southern hemisphere). A large amount of latent heat energy is released into the atmosphere generating strong winds.

It is well known that a hurricane has an eye which may range from 10 to 60 km in diameter within which there is no rain, almost no clouds, and extremely diminished wind. Hurricanes move to the west at a speed of 10–25 km/h. Due to the Coriolis effect, the westward hurricane path usually deviates toward NW in the Northern hemisphere. However, hurricanes are usually steered by high-level winds and their speed and direction often become erratic. When hurricanes move over cold water or land, they quickly loose energy and release moisture as intense rainfall.

Category	Wind speed		Central pressure (Millibar)	Damage
	(mph)	(m/s)		
1	74–95	33–42	≥980	Minimal
2	96–110	43–49	979–965	Moderate
3	111–130	50–58	964–945	Extensive
4	131–155	59–69	944–920	Extreme
5	≥155	≥70	≤920	Catastrophic

Table 1.1 *The Saffir–Simpson hurricane damage potential scale*

1.1.2 Mean wind speed

One of the most important characteristics of wind is its speed; in particular, the speed in its direction of advancement, called longitudinal velocity, the magnitude of which is critical for the design and safe operation of marine systems. In order to present the magnitude of wind speed, we may consider the average value and the severity of the fluctuating components called turbulence. The former is discussed in this section, while the latter will be discussed after the wind spectrum is clarified.

Mean wind speed is shown in the sample given in Figure 1.1. The mean speed is obtained most commonly in a 10 min recording; however, this is not always the case. For instance, in the measurements by the NOAA buoys the wind speed is averaged over an 8 min observation. The mean wind speed is a function of height above the sea level. The mean wind speed at height z above the sea level, denoted by $\bar{U}_z$, is commonly evaluated by the following logarithmic law given by

$$\bar{U}_z = \frac{u_*}{K}\ln(z/z_0) \tag{1.1-1}$$

where u_* is shear (friction) velocity (m/s) $= \sqrt{\tau/\rho}$, τ surface stress, ρ air density, K von Karman constant, 0.4, and z_0 roughness length.

The value of the roughness length depends on the intensity of friction between the air and sea surface. The appropriate value to be considered for a hurricane-generated sea surface will be discussed later. At this stage, we may evaluate the mean wind speed, $\bar{U}_z$, without knowledge of the roughness length, i.e. let $\bar{U}_{10}$ be the mean wind speed at 10 m height above the sea level which is the reference height most commonly considered for wind speed. Then, by writing $\bar{U}_{10}$ in the form given in Equation (1.1-1), the mean wind speed at height z above the sea level can be written in terms of the mean wind speed at 10 m height as follows

$$\bar{U}_z = \bar{U}_{10} + 2.5u_* \ln(z/10) \tag{1.1-2}$$

where $\bar{U}_{10}$ is mean wind speed at 10 m height in m/s.

The shear velocity, u_*, is a function of the height above the sea surface and is given by

$$u_* = \sqrt{\tau/\rho} = \sqrt{C_z}\bar{U}_z \tag{1.1-3}$$

where C_z is defined as the surface drag coefficient evaluated from wind measurement at height z. For convenience, we may consider the drag coefficient evaluated at the reference height of 10 m to be denoted by C_{10}. Many studies have been carried out on the surface drag coefficient C_{10} based on data obtained for a wide range of wind velocities over the seaway.

Figure 1.2 shows the results of some typical important studies on the surface drag coefficient. As seen in the figure, the coefficient can be presented as a linear function of the mean wind speed at 10 m height. In particular, Garrat (1977) makes an extensive review of studies and derives the following formula

$$C_{10} = (0.75 + 0.067\bar{U}_{10}) \times 10^{-3} \tag{1.1-4}$$

On the other hand, Wu (1980, 1982) developed the following formula based on data including hurricane winds

$$C_{10} = (0.8 + 0.065\bar{U}_{10}) \times 10^{-3} \tag{1.1-5}$$

The values of the drag coefficient given in Equations (1.1-4) and (1.1-5) are almost the same as seen in Figure 1.2.

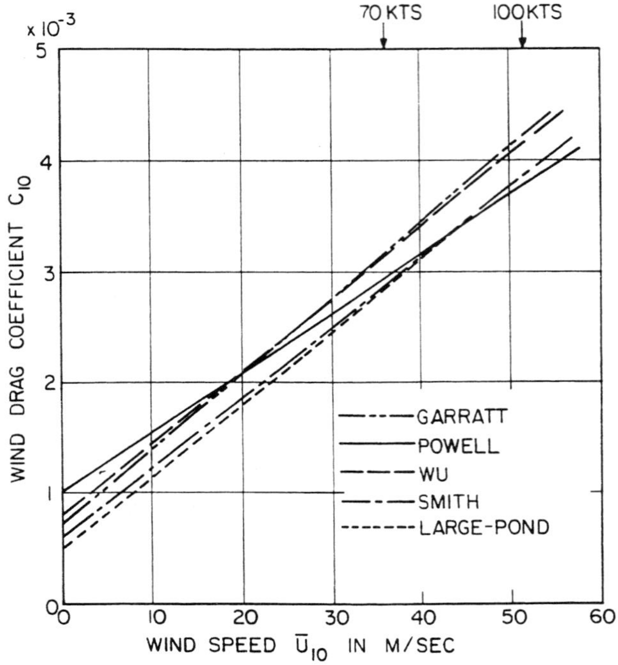

Figure 1.2 – Wind drag coefficient C_{10} as a function of mean wind speed U_{10} (Ochi and Shin, 1988).

Figure 1.3 shows the mean wind speed at various heights above the sea level for a given wind speed at 10 m height using the drag coefficient formulation given in Equation (1.1-5).

In regard to the roughness length z_0, it can be evaluated in terms of the surface drag coefficient C_z; i.e. from Equation (1.1-1), z_0 can be written as

$$z_0 = z\, e^{-K\bar{U}_z/u_*} \tag{1.1-6}$$

while from Equation (1.1-3), we have $\bar{U}_z/u_* = 1/\sqrt{C_z}$. Hence, z_0 becomes

$$z_0 = z\, e^{-K/\sqrt{C_z}} = z\, e^{-0.4/\sqrt{C_z}} \tag{1.1-7}$$

On the other hand, Charnock (1955) finds the following relationship from his analysis of wind stress data over the water surface of a reservoir

$$z_0/(u_*^2/g) = \text{constant} \tag{1.1-8}$$

Garrat (1977) gives the Charnock constant in Equation (1.1-8) as 0.0144 for a stable atmosphere over the ocean, while Wu (1980) obtains a value of 0.0185 from his analysis of observations at sea for wind speed greater than 5 m/s at 10 m level.

Let us evaluate the value of the roughness length for hurricane KATE with a measured wind speed of 47.3 m/s at 10 m height above the sea level. We have $C_{10} = 3.875 \times 10^{-3}$ from Equation (1.1-5). By using these values, we have $z_0 = 0.0162$ m from Equation (1.1-7). If we evaluate z_0 by Equation (1.1-8) using Wu's constant 0.0185 and a shear stress $u_* = 2.95$ m/s at 10 m height computed by Equation (1.1-3), we have $z_0 = 0.0164$ m. Thus, the roughness length for hurricane KATE is obtained as 0.0163 m. In view of the available roughness length

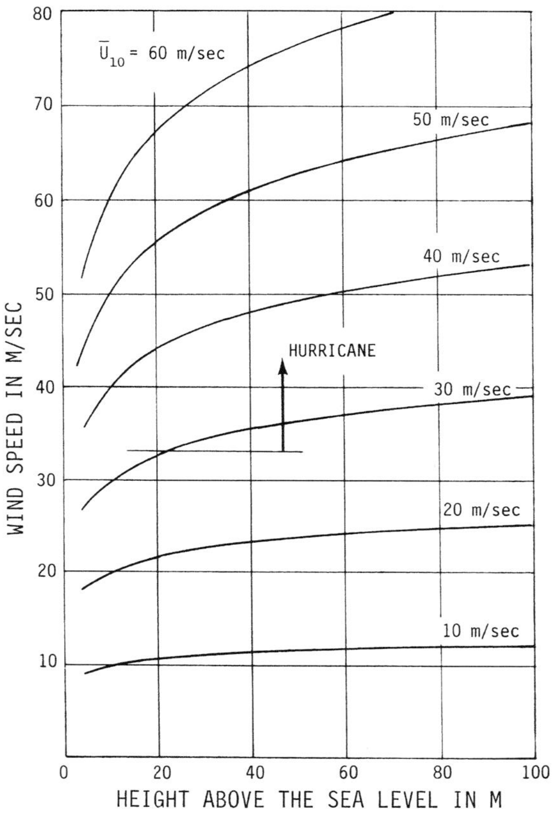

Figure 1.3 – Mean wind speeds at various heights above sea level for a given wind speed at 10 m height.

in storm seas being approximately 0.01 m or greater, the computed value appears to be very reasonable for hurricane-generated seas.

Information on various wind characteristics is generally referred to as that at 10 m height above the sea level. For presenting the wind characteristics at lower heights, SethuRamen and Raynor (1975) considered the roughness Reynolds number (RRN) defined as $u_* z_0 / \nu$, where u_* is the surface friction velocity, z_0 the roughness length and ν the kinematic viscosity. They designate the sea surface to be moderately rough for $0.15 \leq \text{RRN} \leq 4.0$, and fully rough if $\text{RRN} > 4.0$. They evaluate the surface drag coefficient at 6 m height in a fully rough sea surface condition as 1.9×10^{-3} in the mean wind speed range from 3 to 10 m/s; this value being much greater than that at 10 m height shown in Figure 1.2. Further, they find Charnock's constant given in Equation (1.1-8) as 0.072 at 6 m height in a fully rough sea surface condition which is also very large in comparison with the value at 10 m height.

1.1.3 Turbulent wind spectrum

As shown in Figure 1.1, the time history of wind speed shows that it can be decomposed into two components; a constant component and a turbulence component having a variety of frequency elements. The energy associated with fluctuating turbulent wind speeds is represented by an energy spectrum which is a function of the severity of the mean wind speed, and thereby the height above the sea level and the shear velocity. The spectral density function and the frequency of the turbulent winds are commonly presented in the following dimensionless form

$$S(f_*) = f\,S(f)/u_*^2 \tag{1.1-9}$$

$$f_* = fz/\bar{U}_z \tag{1.1-10}$$

where f_* is the dimensionless frequency, f frequency (Hz), z height above sea level (m), $\bar{U}_z$ mean wind speed at height z (m/s), $S(f)$ spectral density function of wind speed (m^2/s), and u_* shear velocity (m/s).

In general, three components (longitudinal, horizontal, and vertical) of the spectrum are considered for the turbulent winds. However, the longitudinal component is by far the most significant for the design and operation of marine systems; hence, we may consider only the longitudinal component spectrum in the following.

Many mathematical formulations representing the spectral density function of turbulent winds have been developed. However, almost all these formulations are based on wind data obtained on land for evaluation of the response of tall buildings and towers to turbulent winds. As stated earlier, turbulent wind energy spectra over a seaway contain significantly more energy than over-land spectra at low frequencies because of the wide open space over a seaway. This was found earlier by Busch and Panofsky (1968), but had not been considered to be a serious problem until the significance of turbulent wind was recognized for operation of particular types of ocean platforms such as the tension-leg platform.

In order to evaluate the response of marine systems to wind force, Ochi and Shin (1988) review turbulent wind spectra obtained from data measured over a seaway at various geographical locations. Table 1.2 summarizes the measurements of wind over sea by many researchers, while Figure 1.4 compiles the dimensionless spectral density functions. As shown in Table 1.2, the data covers a variety of wind speeds, heights above the water surface

	Location	Height above sea surface (m)	Mean wind speed (m/s)
Eidsvik (1985)	Norwegian sea	110	2.0–36.0
Schmitt, Friehe, and Gibson (1979)	North Pacific	29	5.5–11.8
Miyake *et al* (1970)	400 m offshore, Vancouver Canada	1.4–4.5	3.8–8.6
Takeda (1981)	1 km offshore, Kanazawa Japan	0.5–5.1	≤7
Pond *et al* (1971)	Barbados, Caribbean	8	3.9–7.2
Weiler and Burling (1967)	400 m offshore, Vancouver Canada	1.7–2.7	1.4–10.0
Smith (1967)	Nova Scotia, Canada	1.6–4.2	6.0–22.0

Table 1.2 *Measurements of turbulent winds over the sea (Ochi & Shin, 1988)*

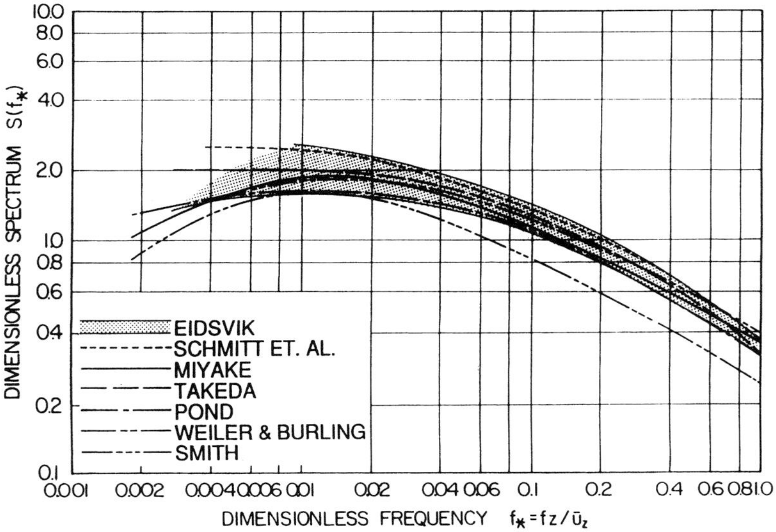

Figure 1.4 – Comparison of turbulent wind spectra obtained from measured data over a seaway (Ochi and Shin, 1988).

and geographical locations. Nevertheless, there is little scatter in the magnitude of the dimensionless spectral density functions except at very low frequencies.

Specific note should be given to Eidsvik's work (Eidsvik, 1985). He carries out an extensive analysis of turbulent wind spectra on a total of 3660 wind speed time histories covering a mean wind speed range from 2 to 36 m/s at 110 m height above the water surface. He classifies the data into 15 wind speed groups and generated 15 spectra in sequence of increasing wind severity. However, only the largest and smallest wind spectra are shown in Figure 1.4. In presenting Eidsvik's data in dimensionless form, the value of the drag coefficient C_{10} is maintained constant, 1.5×10^{-3}, for all wind speeds, since this value was estimated through his measurements. It is of interest to observe that Eidsvik' upper and lower bound spectra encompass most of other data obtained elsewhere.

We may draw the average curve of all measured spectra (except for Smith's results) shown in Figure 1.4, the result of which is shown in Figure 1.5 along with mathematical formulations representing the average curve. Prior to explanation of these mathematical formulations, it may be well to discuss the general characteristics of the turbulent wind spectral density function here.

Turbulent energy is generated in large eddies at low frequencies and dissipates in small eddies at high frequencies. In the intermediate range (called the inertial subrange), the turbulent energy production is balanced by dissipation, and thereby the turbulence is nearly isotropic. It is expected that in this range the kinematic energy is determined solely by the rate of dissipation (called the Kolmogorov law) given in terms of wave number as follows

$$S(k) = \alpha \varepsilon^{2/3} k^{-5/3} \tag{1.1-11}$$

where ε is dissipation rate and k wave number.

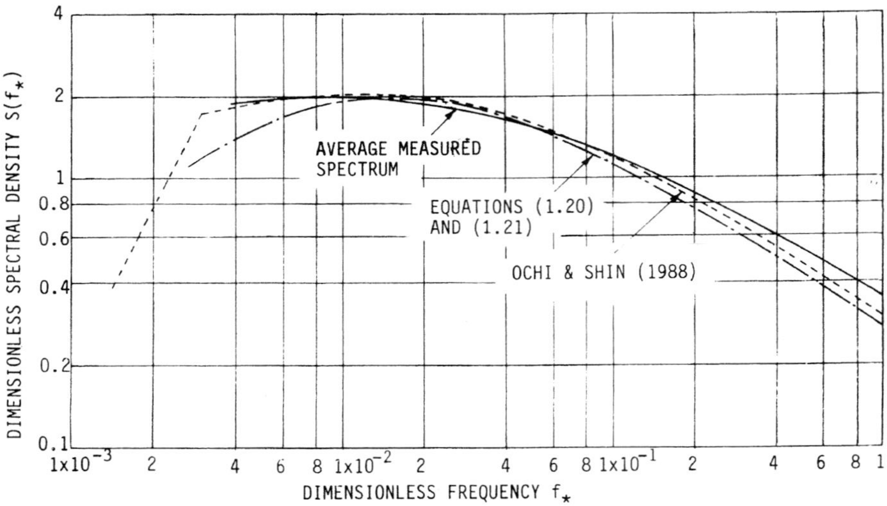

Figure 1.5 – Mathematical presentation of average measured wind spectra over a seaway. Equations (1.20) and (1.21) in the figure stand for Equations (1.1-20) and (1.1-21), respectively, in the text.

By applying Taylor's hypothesis, the wave number may be written as

$$k = (2\pi f)U^{-1} \tag{1.1-12}$$

Then, we may write Equation (1.1-11) in terms of the frequency f as

$$S(f) = \alpha\varepsilon^{2/3}(2\pi U^{-1})^{-2/3}f^{-5/3} \tag{1.1-13}$$

Further, by using the relationship given in Equations (1.1-9) and (1.1-10), we may write in dimensionless form

$$S(f_*) = \frac{\alpha\varepsilon^{2/3}}{u_*^2}\left(\frac{z}{2\pi}\right)^{2/3} f_*^{-2/3} \tag{1.1-14}$$

Equation (1.1-14) is often written in terms of dimensionless dissipation rate as follows

$$S(f_*) = \frac{\alpha}{(2\pi K)^{2/3}}\phi_\varepsilon^{2/3} f_*^{-2/3} \tag{1.1-15}$$

where ϕ_ε is dimensionless dissipation rate $= \varepsilon Kz/u_*^3$ and K von Karman constant, 0.4.

Thus, following the Kolmogorov law it is obtained that the dimensionless turbulent spectral density function in the inertial subrange is proportional to $f_*^{-2/3}$. The inertial subrange is approximately $f_* \geq 0.1$ which is not a particularly significant range for the design of marine systems. Nevertheless, it may be appropriate to express the average spectral density function obtained from data over the seaway such that the requisite condition discussed above is satisfied.

Considering the conditions given in Equations (1.1-14) and (1.1-15), the dimensionless spectral density function of almost all turbulent winds is presented in the form of

$$S(f_*) = \frac{Af_*}{(1 + Bf_*)^{5/3}} \tag{1.1-16}$$

or

$$S(f_*) = \frac{Af_*}{(1 + Bf_*^2)^{5/6}} \tag{1.1-17}$$

These formulae result in the spectral density being proportional to $f_*^{-2/3}$ in dimensionless form and proportional to $f^{-5/3}$ in dimensional form over the inertial subrange.

A more general expression of the dimensionless turbulent wind spectral density function is given by Olesen, Larsen and Hojstrup (1984) as

$$S(f_*) = \frac{Af_*^{\gamma}}{(1 + Bf_*^{\alpha})^{\beta}} \tag{1.1-18}$$

Tieleman(1995) also discusses the general expression for the spectral density formulation in detail. For the expression given in Equation (1.1-18), the following condition is required to satisfy the Kolmogorov law

$$\gamma - \alpha\beta = -2/3 \tag{1.1-19}$$

With the aid of the above equation, the average line of the dimensionless spectral density function obtained from data measured over the seaway is presented by the following form and a comparison with the average curve is shown in Figure 1.5.

$$S(f_*) = \frac{613f_*}{(1 + 100f_*)^{5/3}} \tag{1.1-20}$$

As seen, Equation (1.1-20) represents well the measured data in the dimensionless frequency range from 0.004 to 0.1. Considering a hurricane wind speed of 50 m/s at 10 m height above the sea surface, this dimensionless frequency range is equivalent to a period range from 50 to 2 s, which sufficiently covers the significant period range for design of marine systems.

However, when the dimensionless spectral density function given in Equation (1.1-20) is converted to the dimensional spectrum, $S(f)$, the value of the spectral density at $f = 0$ is finite. In order to avoid this contradiction, we may consider another formulation given by

$$S(f_*) = \frac{1150f_*^{1.10}}{(1 + 110f_*)^{1.77}} \tag{1.1-21}$$

Although Equation (1.1-21) results in the spectral density $S(f)$ being zero for the frequency $f = 0$, the difference between Equations (1.1-20) and (1.1-21) is less than 1% in the range where the frequency f_* is between 0.004 and 0.1. Since the difference in the magnitude of the spectral density functions given in Equations (1.1-20) and (1.1-21) is extremely small, and since it is known that the very low frequency domain is unstable, it appears that the simple formulation given by Equation (1.1-20) can be appropriately considered for a hurricane-associated wind spectrum over the seaway.

The dimensional spectral density function of Equation (1.1-20) is given by

$$S(f) = \frac{613u_*^2}{\{1 + 100(fz/\bar{U}_z)\}^{5/3}}(z/\bar{U}_z) \tag{1.1-22}$$

In order to present the average curve more precisely Ochi and Shin (1988) develop the following approximate formulation

$$S(f_*) = \begin{cases} 1.75(f_*/0.003)^2 & \text{for } f_* \leq 0.003 \\ \dfrac{420 f_*^{0.70}}{(1 + f_*^{0.35})^{11.5}} & \text{for } 0.003 \leq f_* \leq 0.1 \\ \dfrac{650 f_*}{(1 + 100 f_*)^{5/3}} & \text{for } f_* \geq 0.1 \end{cases} \tag{1.1-23}$$

The high frequency part of the Ochi and Shin's original formula is modified in Equation (1.1-23) in order to satisfy the Kolmogorov law. Equation (1.1-23) is also included in Figure 1.5.

Lumley and Panofsky (1964) state that data have shown that the variance of turbulent winds at any height has a functional relationship with the mean wind speed at that height and this results in the standard deviation, σ, being proportional to the shear velocity u_*. In order to examine whether or not this property is applicable to hurricane winds over the seaway, computations are made for two wind speeds at various heights above the sea level by using Equation (1.1-20); one for a wind speed of 40 m/s, the other 50 m/s, both at a 10 m height.

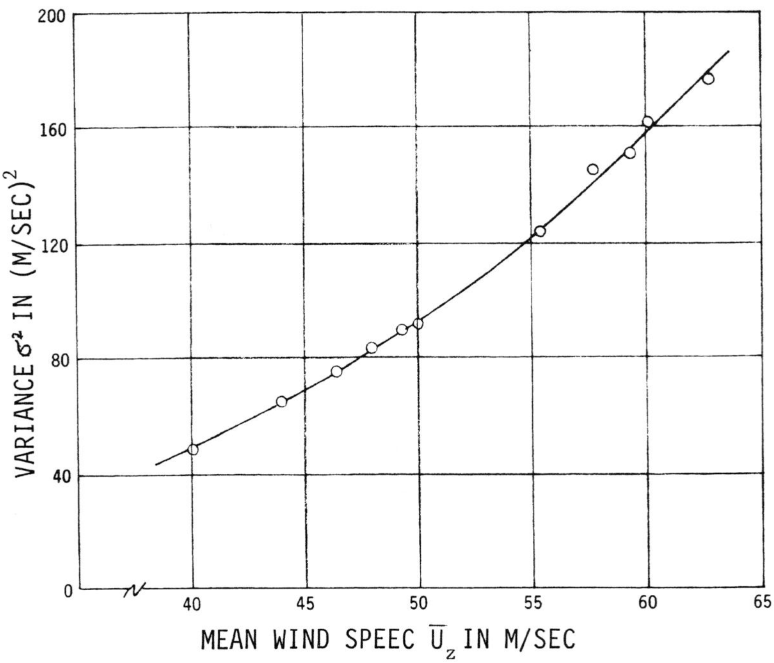

Figure 1.6 – Computed variances of turbulent wind speed plotted as a function of mean wind speed U_{10}.

It is noted that in order to evaluate the variance, Equation (1.1-20) should be integrated from 0 to ∞. Kaimal (1973), however, suggests that for more reliable results an integration range of $0.005 \leq f \leq 10$ Hz should be considered eliminating the very low and high frequencies. In the present computation, the range of integration is taken over $0 \leq f \leq 10$ Hz.

Figure 1.6 shows the computed variances plotted against the mean wind speed. As seen, the computed result agrees with the findings of Lumley and Panofsky; there is a functional relationship between them irrespective of the height above the sea level. If we plot the standard deviations against shear velocity, there is a linear functional relationship between them as shown in Figure 1.7. In the case of hurricane winds, the relationship is given by

$$\sigma = 3.0u_* \tag{1.1-24}$$

The constant involved in Equation (1.1-24) is larger than the 2.5 commonly known for winds, but the difference may be attributed to the magnitude of hurricane wind speeds used in the computations.

It may be of interest to compare the average turbulent wind spectrum obtained from measured data over the seaway (Equation 1.1-20) with typical formulations considered in

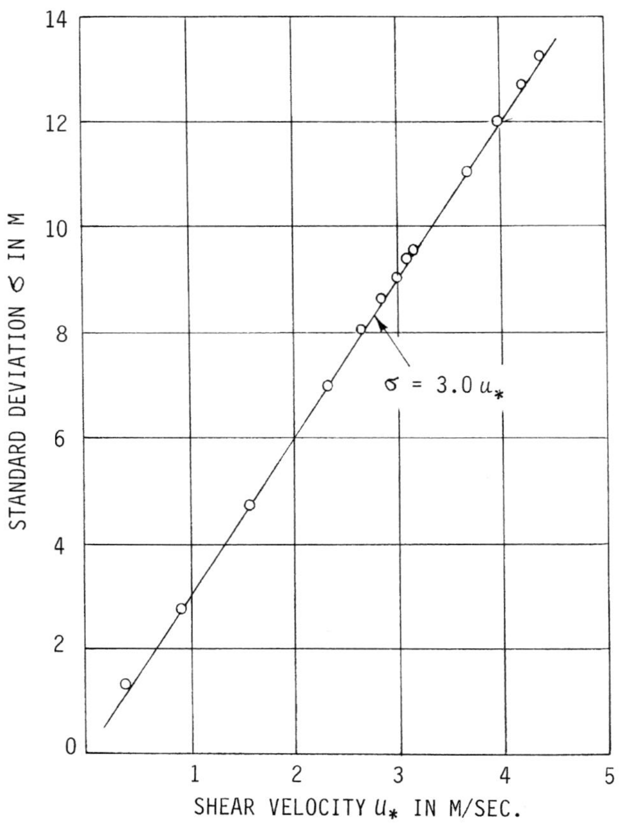

Figure 1.7 – Standard deviation of turbulent wind speed plotted as a function of shear velocity u_*.

engineering design. These include Davenport (1967), Miyake *et al* (1970), Kaimal *et al* (1972), Takeda (1981), Simiu and Leigh (1983) and Kareem (1985), among others. Some of these spectra are developed based on data measured over land in order to provide information for vibration problems of buildings and towers having very high natural vibration frequencies.

Figure 1.8 shows a comparison between the average measured turbulent wind spectrum over a seaway and the spectral formulations of Davenport, Kaimal *et al* and Kareem. Computations are made in dimensional form of wind spectra at 10 m height above sea level for a mean wind speed of 47.3 m/s (hurricane KATE's wind speed). As seen, the average spectral formulation representing measured data over a seaway is greater than any of the spectral formulations for low frequencies. The difference becomes pronounced for frequencies lower than 0.03 which is equivalent to a period of 33 s or greater, a critical range of periods for some offshore structures (tension-leg platform, for example) and moored vessels in a seaway.

In general, the magnitude of lower frequency turbulent wind energy is unstable resulting in the spectral density being scattered when presented in dimensionless form. In order to deal with this problem, Kaimal *et al* (1972) present the spectral density function by a single line in the inertial subrange spreading into several branches at low frequencies as a function of dimensionless height, z/L, since the dimensionless dissipation rate given in Equation (1.1-15) may be expressed by a function of z/L. (see also Leavitte (1975)). Here, L is

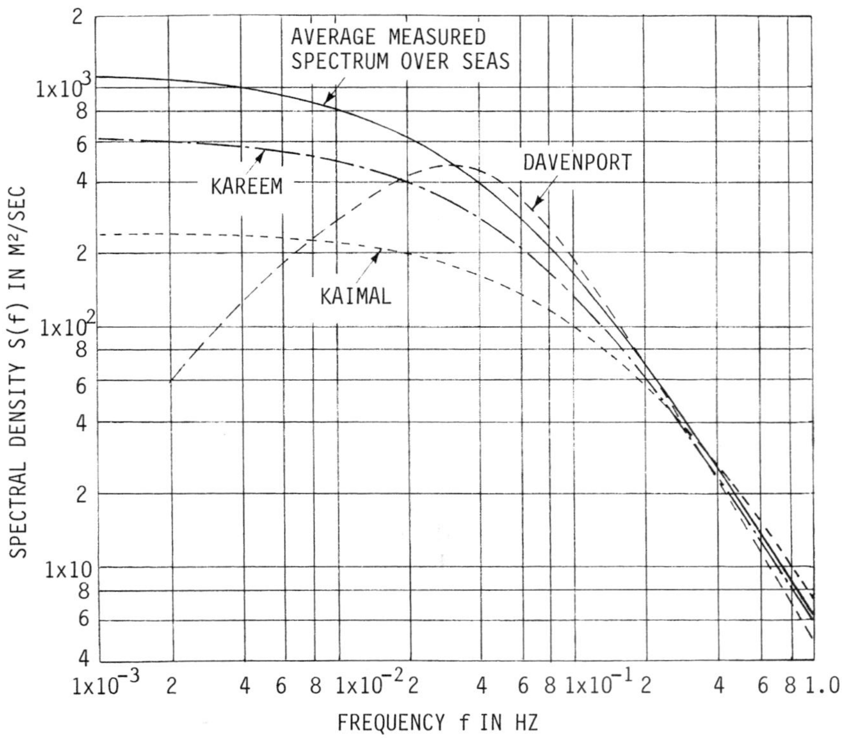

Figure 1.8 – Comparison of various wind spectral formulations computed for mean wind speed of 47.3 m/s at 10 m height above sea level.

called the Monin–Obukhov length scale and z/L is a measure of the relative significance of heat convection and mechanical turbulence. Schmitt, Friehe, and Gibson (1979) carry out a similar analysis on data obtained over the ocean. Their results show that the lower frequency range of the vertical wind velocity spectrum agrees with the trend obtained by Kaimal *et al*, but the agreement is poor for the longitudinal wind velocity spectrum.

1.1.4 Gusts

Turbulent winds are fluctuating around the mean wind speed, and their severity is represented by the energy of the spectral density function. In particular, the following turbulent intensity at height z above the sea level, denoted by $I(z)$, is commonly used as a criterion for representing the turbulent severity

$$I(z) = \frac{\sigma_z}{\bar{U}_z} \tag{1.1-25}$$

where σ_z is standard deviation, the square root of the area under the spectrum.

We may write the turbulent intensity from Equations (1.1-1) and (1.1-24) with $K = 0.4$ as follows

$$I(z) = \frac{1.2}{\ln(z/z_0)} \tag{1.1-26}$$

where z_0 can be evaluated from Equation (1.1-8).

If we compute the turbulent intensity for hurricane KATE with $\bar{U}_{10} = 47.3$ m/s, we have the variance $\sigma^2 = 77.78\ \text{m}^2$ (see Figure 1.6). Hence, from Equation (1.1-25) the turbulent intensity can be evaluated as $\sigma/\bar{U}_{10} = 0.186$ at 10 m height. The same result can be obtained from Equation (1.1-26) by using $z_0 = 0.0163$ as evaluated in Section 1.1.2.

Unfortunately, only a limited amount of measured data on the turbulent intensity of hurricane winds are available. Schroeder, Smith and Peterson (1998) evaluate the longitudinal turbulent intensity for each 10 min segment during hurricane BOB which passed 55 km off the measurement site. They obtain the maximum and average turbulent intensities as 0.169 and 0.146, respectively, at 19 m height. From measured data during typhoons, Choi (1978) gives 0.175 as the average value of turbulent intensity at 10 m height. Ishizaki (1983) evaluates the turbulent intensities of typhoon winds as a function of mean wind speed up to 30 m/s and shows that the intensity reduces with increase of mean wind speed. In particular, the turbulent intensity is very high (on the order of 0.3–0.4) for a mean wind speed of about 5 m/s. On the other hand, for mean wind speeds over 40 m/s, though the number of observations is small, the turbulent intensity is low and is almost constant.

Figure 1.9 shows the turbulent intensities evaluated from the variance shown in Figure 1.6 for hurricane wind speeds over 40 m/s. As seen, the turbulent intensity increases very slowly with increase in mean wind speed. The trend of increase in intensity with mean wind speed appears to be reasonable, since from Equations (1.1-3), (1.1-24) and (1.1-25) we have

$$I(z) = \sigma/\bar{U}_z = 3.0\sqrt{C_z} \tag{1.1-27}$$

where the drag coefficient C_z is an increasing function of height. Hence, $I(z)$ increases with increase in height z.

Another parameter often considered for indicating the gust severity is the gust factor defined as the ratio of the maximum wind speed of a given period to the mean wind speed observed in a specified time. For the NOAA buoy data the maximum turbulence having a 5 s period in 8 min observation is defined as a gust. However, the definition of the gust factor is

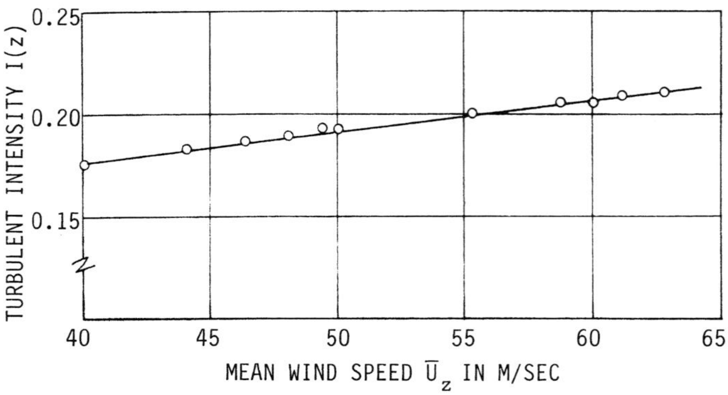

Figure 1.9 – Turbulent intensity computed with variance shown in Figure 1.6 plotted as a function of mean wind speed.

not consistent. Some define the gust factor as the ratio of the maximum turbulence having a 3 s period (Shears and Bell, 1979), while others consider the maximum turbulence having a 4.5 s period (Ishizaki, 1983) to the mean wind speed in 1 h observation.

On the other hand, Durst (1960), Krayer and Marshall (1992) and Schroeder and Smith (1999) define the gust factor as the ratio between the peak wind speed of given duration within a particular data segment and the mean wind speed of the same segment. By this definition, the magnitude of the gust factor depends on the length of the specified time duration; the gust factor decreasing with increase in time duration of the segment. In particular, Krayer and Marshall analyze hurricanes FREDERICK, ALICIA, ELENA and HUGO data and evaluate the gust factor for the maximum gust having a 2 s period in a 10 min segment. They show an average gust factor of 1.55 in 1 h observation. If the data are analyzed in 20 min segments, the average gust factor reduces to 1.48.

In the following, let us evaluate the gust factor for hurricane winds from the NOAA buoy data. Figure 1.10 shows the gust wind speed obtained from Buoy 42003 at 10 m height during hurricane KATE and that obtained from Buoy 410202 during hurricane GLORIA. GLORIA's wind speed measured at 5 m height is converted to that at 10 m height in the figure for comparison. By drawing the average line for each set of data, a gust factor of 1.23 is obtained for KATE and 1.36 for GLORIA. The reason for this difference is not clearly known. There was a difference in winds between these two hurricanes in that winds of moderate intensity were blowing continuously for more than 10 days before hurricane GLORIA came to the site.

Since the magnitude of turbulent winds is randomly changing, it is certainly an expedient approach to evaluate the extreme magnitude which is expected to occur in a specified time period. Counihan (1975) examines the probability distribution of turbulent velocity and concluded that the velocity can be considered as a Gaussian random process. If this is the case, it may be possible to evaluate extreme gust magnitudes by applying the techniques of the stochastic process approach as commonly employed in predicting extreme heights of ocean waves. For example, Davenport (1964), Forristall (1988) apply the concept of extreme values for estimating the maximum gust.

However, there is some reservation in the application of the currently available extreme value prediction method to turbulent winds in that the time history of turbulent fluctuation shows it does not cross the mean wind speed line; instead, it crosses the average

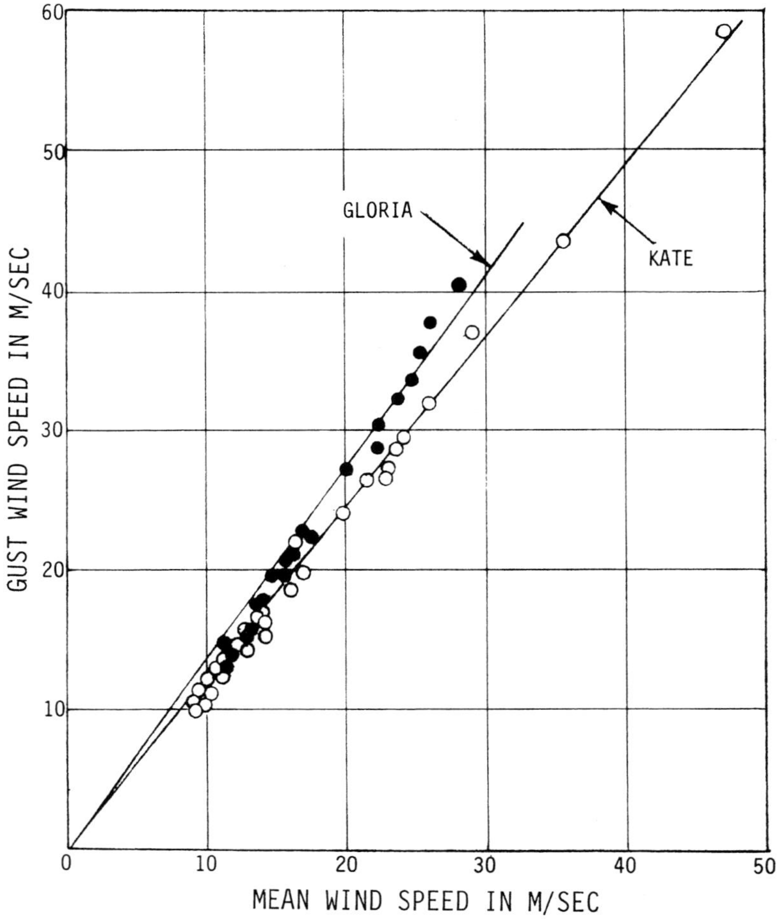

Figure 1.10 – Gust wind speeds plotted against the mean wind speeds obtained in hurricanes KATE and GLORIA.

line of a low frequency (or frequencies). For instance, in the example of the time history of wind shown in Figure 1.1, turbulence is crossing the average line of the low frequency component of approximately 0.125 Hz (8 s period) and, in addition, this low frequency component is crossing the average line of an even lower frequency component of approximately 0.035 Hz (28 s period). At present, it is not possible to evaluate the extreme value of turbulent winds under this situation. Development of a new method through the stochastic process approach is highly desirable.

1.2 Sea state generated by hurricane winds

Wind-generated storm seas are referred to, in general, as seas generated by quasi steady-state winds blowing for several hours, sometimes several days, in a certain direction over a relatively long fetch in the open ocean. Under this situation, the sea surface becomes highly irregular and non-repeatable; hence, deterministic description of the sea surface is not possible in time and space. The severity of the sea, however, may be addressed by a wave spectrum representing the time average wave energy, i.e. with winds blowing for several hours, wave energy originates at the higher frequencies of the spectrum, and then the wave spectrum slowly increases in magnitude with time by adding energy at the lower frequencies. Concurrently, the modal frequency, defined as the frequency where the spectrum peaks, migrates slowly to the lower frequencies. This description of the growth of sea state is germane to ordinary wind-generated seas irrespective of their severity. However, this is not the case for sea states generated by hurricanes.

For hurricane-generated seas, the input source of energy which generates random waves is advancing with a certain speed usually in the same direction as the wave energy is progressing. Therefore, the wind speed at a certain location in the ocean in the path of a hurricane increases very fast and the time duration of a given wind speed is much shorter than that observed during an ordinary storm. Nevertheless, the severity of the sea (significant wave height) increases very fast. In particular, the sea state increases rapidly when the hurricane center approaches within a distance of approximately 150 km (95 miles) from the site. Consequently, several interesting features which are quite different from those observed in ordinary storms can be recognized in the shape of wave spectra of hurricane-generated seas. This will be discussed in detail in Section 2.1.

Many studies have been carried out on the severity of sea states generated by hurricane winds from various view points and through different approaches. Among others, hindcasting and forecasting techniques for estimating hurricane-generated seas have recently made significant progress. The numerical prediction models certainly provide valuable information on the overall mapping of sea states generated by specified hurricane winds, although more confirmation of the techniques through comparison with measured data is highly desirable. The readers may refer to the publications authored by SWAMP Group (1985), WAMDI Group (1988) and Komen *et al* (1994).

This section addresses the results of studies on the deterministic relationship between hurricane wind speed and sea severity obtained from analysis of measured data. The stochastic evaluation of sea severity through wave spectra is the fundamental approach employed in this text, and will be discussed in detail in Chapter 2 and thereafter.

It is commonly known that the severity of the sea generated by hurricane winds depends on the magnitude of (a) maximum sustained wind speed, (b) minimum surface pressure in the hurricane eye, (c) forward velocity of the hurricane center, and (d) radius of maximum wind speed. There is no doubt that each of these parameters contributes to the generation of a severe sea by hurricane winds. However, for the design and operation of marine systems such as ships and offshore structures and for the environmental safety of nearshore structures, information needed is the severity of the sea (such as significant wave height and wave spectrum) and wind speed at the site. Both the sea severity and wind speed at a specified location along the path of a hurricane increase with time as the hurricane approaches the site. Therefore, it is highly desirable to obtain the functional relationship between the wind speed and sea severity so that the severest sea condition at the site may be estimated based on a continuous observation of wind speed.

The results of analysis of measured hurricane wind speed and sea severity indicate that there exists a functional relationship between wind speed and sea severity (significant wave height), but the sea severity at the site depends on the sea condition during a week or 10 days prior to the hurricane arrival. This subject will be discussed in detail in the following.

First, let us examine the relationship between wind speed and sea severity under the condition that the sea condition at the site was relatively calm for 7–10 days prior to the arrival of a hurricane. Figure 1.11 shows an example of the relationship between the measured mean wind speed at 10 m height and the significant wave height at 1 h intervals obtained by the NOAA Buoy EB 10 (88.0W, 27.5W) in the Gulf of Mexico (Ochi, 1993). The center of hurricane ELOISE passed within 16 km of the buoy. While the open circles in the figure indicate the wind–wave relationship during the growing stage of the hurricane, the solid circles indicate the relationship during the decaying stage when the hurricane center has already passed the site.

As seen in Figure 1.11, the sea severity increases almost linearly with increase in wind speed during the growing stage. When the wind speed becomes maximum (35.2 m/s) as the hurricane leading edge approaches, the sea severity (significant wave height) becomes maximum (8.8 m). This is followed by the transition stage in which the wind speed rapidly reduces in magnitude (from 35.2 to 8 m/s) during the 2 h of tracking the hurricane center as it passes near the buoy. The sea severity also significantly reduces from 8.8 to 5.0 m during these 2 h. During the next two hours, the hurricane trailing edge passes and the wind speed and sea severity come back to nearly the same levels as they were in the growing stage. This sort of rapid change of sea severity subject directly to wind speed has never been observed in ordinary wind-generated storm seas.

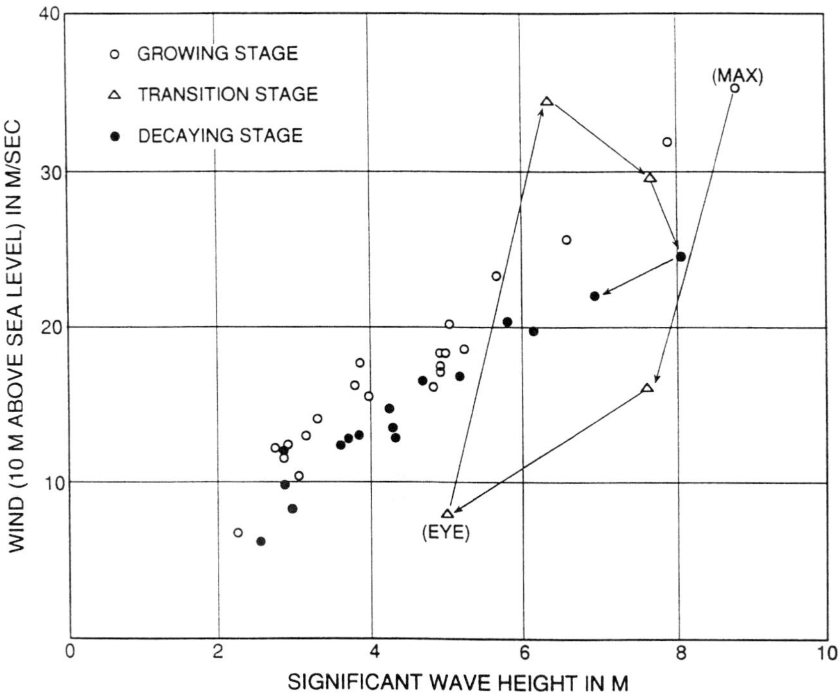

Figure 1.11 – Relationship between mean wind speed and significant wave height obtained by NOAA buoy EB10 during hurricane ELOISE (Ochi, 1993).

As another example of the relationship between wind speed and sea severity, Figure 1.12 shows the relationship obtained during hurricane KATE which passed near the NOAA Buoy 42003 (25.9N, 85.9W) in the Gulf of Mexico. As seen in the figure, the result obtained in hurricane KATE also shows wind speed and sea severity similar to that observed in hurricane ELOISE. The advancing speeds of these two hurricanes were similar; 10.3 knots (5.31 m/s) for ELOISE and 9.2 knots (4.75 m/s) for KATE.

It is noted that the maximum wind speed of hurricane KATE reduced from 47.3 to 16.6 m/s within 1 h when the hurricane center approached the buoy, but the sea severity reduced by only 1 m (10.7–9.8 m). The reason for this small reduction in sea state is not clear. However, it is surmised that if the data were taken at 30 min intervals instead of 1 h, the wind speed as well as sea severity might have shown more reduction when the hurricane center came close to the site.

The sea at the buoy locations prior to hurricanes ELOISE and KATE coming to each site was relatively calm for both cases. In addition, the wind speed and sea severity relationship during the growing stage of these hurricanes is nearly the same. In order to examine whether or not this same relationship holds for other hurricanes with a relatively calm sea condition prior to their arrival, data obtained during the growing stage are analyzed for an additional five hurricanes for which the sea conditions were very mild for at least 7 days before their arrival. These are hurricanes FREDERIC, ANITA, BELLE, CARMEN and CAMILLE, and the results are summarized in Figure 1.13. Data of hurricane CAMILLE were obtained by a resistance-type wave gage attached to a tower located in a water depth of 104 m (Patterson, 1974), while data for all others were obtained by NOAA buoys in deep water

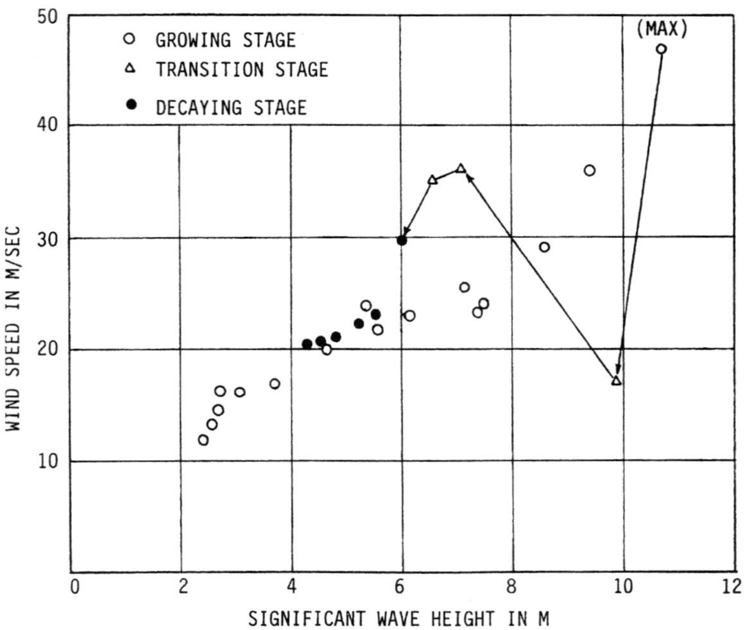

Figure 1.12 – Relationship between mean wind speed and significant wave height obtained by NOAA buoy 42003 during hurricane KATE.

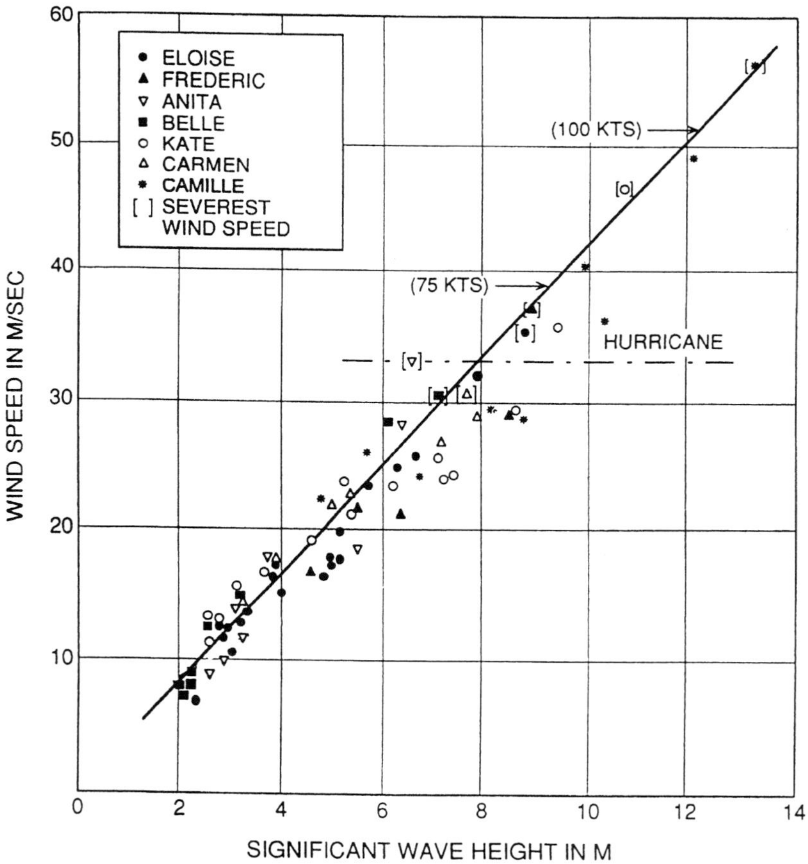

Figure 1.13 – Relationship between mean wind speed and significant wave height obtained during the growing stage of various hurricanes (Ochi, 1993).

(Withee and Johnson, 1975, Johnson and Spear, 1978, Johnson and Renwick, 1978, 1981, and NOAA National Data Buoy Center, 1986). Wind speeds measured at 5 m above the sea level for some buoys are converted to 10 m height. It is noted that the maximum wind speeds of hurricanes BELLE and CARMEN given in the figure are below hurricane level (33.5 m/s). These are wind speeds when the hurricanes passed over or nearby the buoys, but the wind speeds reached hurricane level later.

From the data shown in Figure 1.13, it may be concluded that when a hurricane approaches an area where the sea condition has been relatively calm for 7–10 days, the severity of the sea increases almost linearly with increase in wind speed at the site. By drawing a straight line, which includes the majority of the points in the figure, we can derive the following relationship

$$H_s = 0.24\bar{U}_{10} \tag{1.2-1}$$

where H_s is the significant wave height in meters and $\bar{U}_{10}$ is the mean wind speed at 10 m height in m/s.

Next, let us examine the relationship between wind speed and sea severity for the case when the sea condition has not been calm for 7–10 days before a hurricane comes to the site. For demonstrating the difference in the sea conditions prior to hurricane arrival, Figure 1.14 is prepared. The figure shows the time history of sea conditions (significant wave heights) measured every 1 h for 10 days prior to the severest sea condition measured by buoys for hurricanes KATE and GLORIA. The data on GLORIA were taken by NOAA Buoy 41002 (232.3N, 75.3W) in 1985. As seen in the figure, the sea condition was very mild (significant height on the order of 1–1.3 m) for KATE for almost 1 week before the hurricane center came close to the site, while the significant wave height was 2.5–3 m for GLORIA for the same time frame. This preliminary disturbed sea condition resulted in a much more severe sea condition for GLORIA than for KATE when the hurricane center came to the site.

Figure 1.15(a) and (b) show a comparison of time histories of wind speed and sea severity, respectively, for 2 days prior to the maximum sea state recorded at the buoy locations for hurricanes KATE and GLORIA. GLORIA's wind speeds measured at 5 m height are converted to those at 10 m height in the figure so that a direct comparison with KATE can be done.

As seen in Figure 1.15(a), wind speeds increased rapidly during the last 24 h; in particular, hurricane KATE wind speed increased from 11.0 to 47.0 m/s and this resulted in the magnitude of significant wave height increasing from 2.0 to 10.7 m as shown in Figure 1.15(b). On the other hand, the wind speed of hurricane GLORIA increased from 10.0 to 28.2 m/s in the last 24 h; the increasing rate as well as the magnitude of wind being much less than for KATE. Nevertheless, the magnitude of significant wave height increased from 2.8 to 14.3 m; the sea condition when the hurricane came close to the site being much more severe than for KATE.

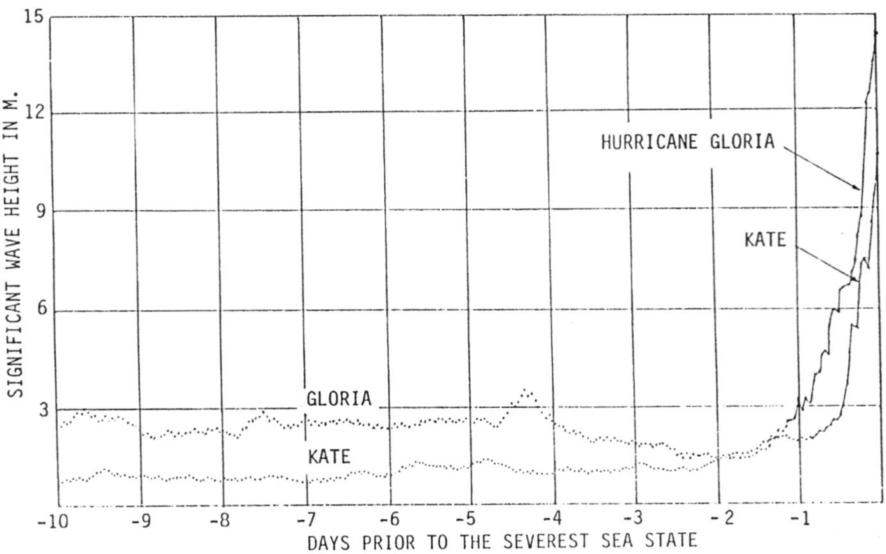

Figure 1.14 – Comparison of time history of sea condition (significant wave height) between hurricanes KATE and GLORIA measured every 1 h during 10 days prior to the severest sea condition.

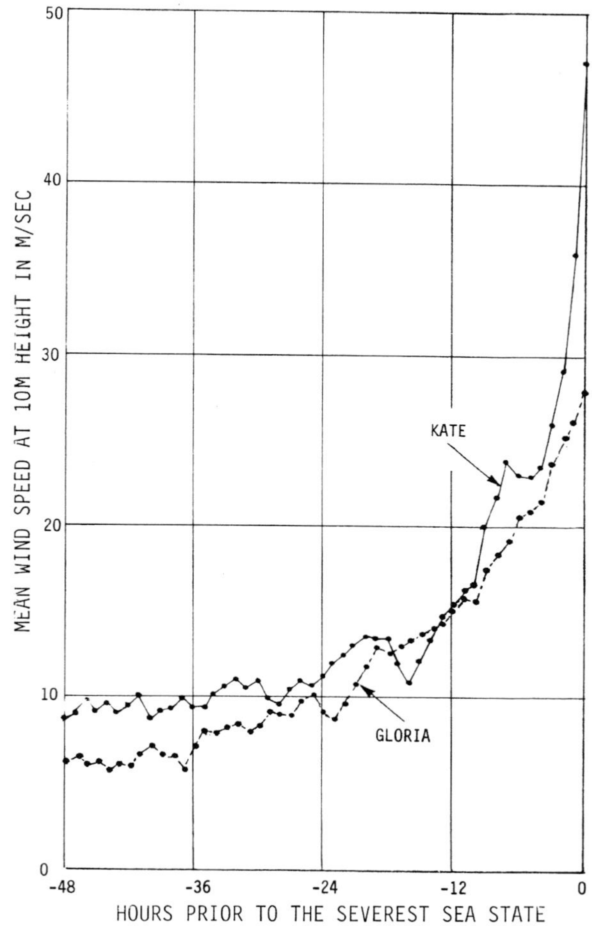

(a)

Figure 1.15 – Comparison of time history of mean wind speed and significant wave height between hurricanes KATE and GLORIA measured for 2 days prior to the severest sea condition. (a) Mean wind speed; (b) significant wave height.

This contradictory relationship between wind speed and sea severity may be attributed to the difference in the sea severity that had existed for some 7–10 days before hurricane came to the site as shown in Figure 1.14. This implies that in the case of hurricane GLORIA, the sea condition had the potential to be easily augmented in severity when the hurricane came to the area.

The relationship between wind speed and sea severity obtained from the buoy data for hurricane GLORIA is given as the solid circles in Figure 1.16. Wind speeds measured at 5 m height for GLORIA are converted to those at 10 m height. The functional relationship may be presented as

$$H_s = 0.078\bar{U}_{10}^{1.57} \qquad (1.2\text{-}2)$$

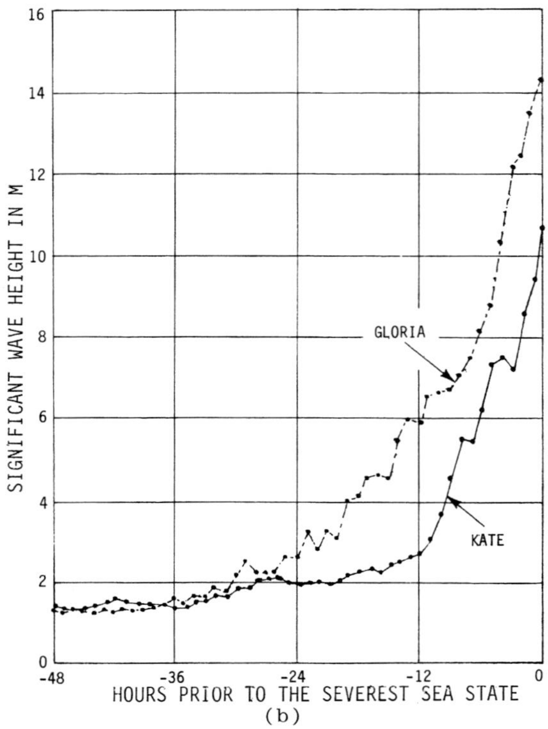

Figure 1.15 (continued).

Included also in Figure 1.16 is Equation (1.2-1) which represents the relationship between wind speed and significant wave height shown in Figure 1.13 for the case of a relatively calm sea prior to the arrival of the hurricane. As seen in Figure 1.16, a significant difference exists in the severity of sea for the same wind speed depending on the prior sea condition. Although no data have been observed so far, functional relationships which fall in between the two lines shown in the figure are expected to occur. In other words, the two lines of significant wave height in the figure are considered to be the limits indicating the functional relationship between the hurricane wind speed and sea severity.

It is noted that the sea severity given by Equation (1.2-2) is perhaps the severest sea state (significant wave height) for a given wind speed. In order to verify this statement, the average wind speed required to generate a specified significant wave height for fully developed seas is also shown in the figure. As can be seen in the figure, the relationship for a fully developed sea is very close to that given by Equation (1.2-2), although a sustained time duration of wind speed is required for the former. The shape of the wave spectrum associated with hurricanes/tropical cyclones and that of fully developed seas are, of course, substantially different since the time duration of a given wind speed is quite different.

It is of interest to compare Equation (1.2-2) shown in Figure 1.16 with that from analysis of data obtained during a strong storm in the North Atlantic (shown by the open circles in the figure) in which winds of 6–14 m/s were blowing for some time before the storm came to Station J (53N, 18W) (Sneider and Chakrabarti, 1973). Waves were measured by

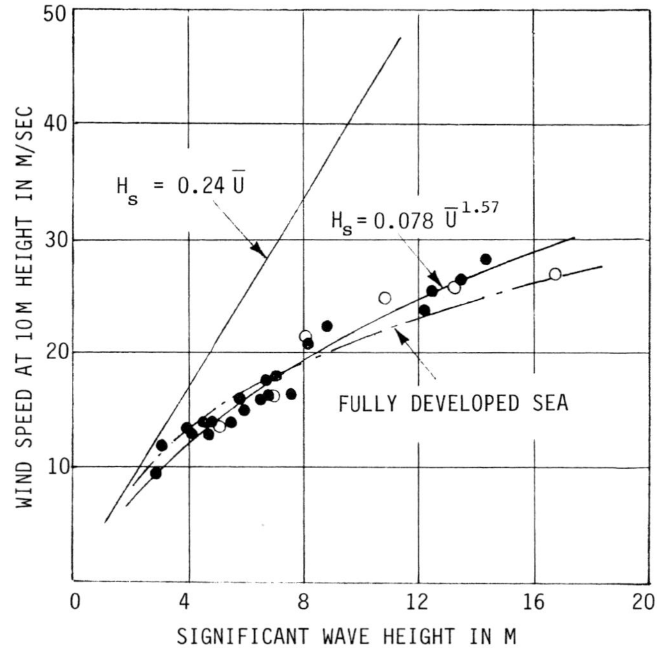

Figure 1.16 – Comparison of wind speed–sea severity relationship obtained from data having different sea severity prior to hurricane arrival to the site; relatively calm sea ($H_s = 0.24\bar{U}$) versus disturbed sea ($H_s = 0.078\bar{U}^{1.57}$).

a Tucker meter installed on the Ocean Weather Ship stationed at that location. When the storm approached the area, the significant wave height increased from 5.0 to 16.8 m in 21 h; while wind speeds increased from 14 to 27.5 m/s as shown in the figure. The significant wave height of 16.8 m is an unusually severe sea state. Since winds were measured on the ship deck which was 19.6 m above the sea level, wind speeds are converted to 10 m height in the figure.

As seen in Figure 1.16, the measured data in this very severe storm in the North Atlantic agree very well with GLORIA's results. The sea condition before the measurements were made was apparently similar to that of hurricane GLORIA. It may, therefore, be safely concluded that a disturbed sea condition (significant wave height on the order of 3 m) prior to the arrival of a heavy storm contributes to the growth rate of sea severity to a great extent.

2

Wave spectra of hurricane-generated seas in deep water

2.1 Features of wave spectra

As briefly discussed in Section 1.2 for hurricane-generated seas, the input source of energy which generates random waves is advancing at a certain speed; thereby, the severity of sea state in the advancing path of the hurricane increases rapidly. In particular, the rate of increase in sea state (significant wave height) at a site is pronounced 3–4 hours prior to the wind speed of the hurricane approaching the site becoming maximum. At this stage, the hurricane center is approximately 150 km (95 miles) from the site. As a consequence, several interesting features in the shape of wave spectra can be observed in hurricane-generated seas which are quite different from those observed in ordinary storms.

One of the features is that the energy of the wave spectrum of hurricane-generated seas is concentrated around the modal frequency. As an example, Figure 2.1 shows a comparison between a wave spectrum computed from data obtained by NOAA Buoy EB-10 (88.0N, 27.5W) during hurricane ELOISE (Withee and Johnson, 1975) and various spectra having nearly the same severity (significant wave height 8.8 m) obtained from data taken in the North Atlantic. As can be seen, the energy density of the spectrum for the hurricane-generated seas is concentrated primarily in the neighborhood of the peak frequency as contrasted with the energy being spread over a wide frequency range, including double peaks, for wave spectra observed during ordinary storms.

The second feature of wave spectra observed during the growing stage of a hurricane is that the location of the modal frequency is almost constant with time, particularly in severe seas. This is pointed out by Young (1998) in his analysis of wave spectra associated with tropical cyclones. Young attributes the consistency of the modal frequency observed during the passage of tropical cyclone ILONA (significant wave height 8.0 m) to the fact that the energy was generated in the intense wind regions of the storm and propagated as swell to the measurement site.

This feature is essentially the same as that observed by Shemdin (1980) who states that waves of dominant energy in the spectrum propagate immediately ahead of the hurricane at the growing stage. This means that the frequency of waves having dominant

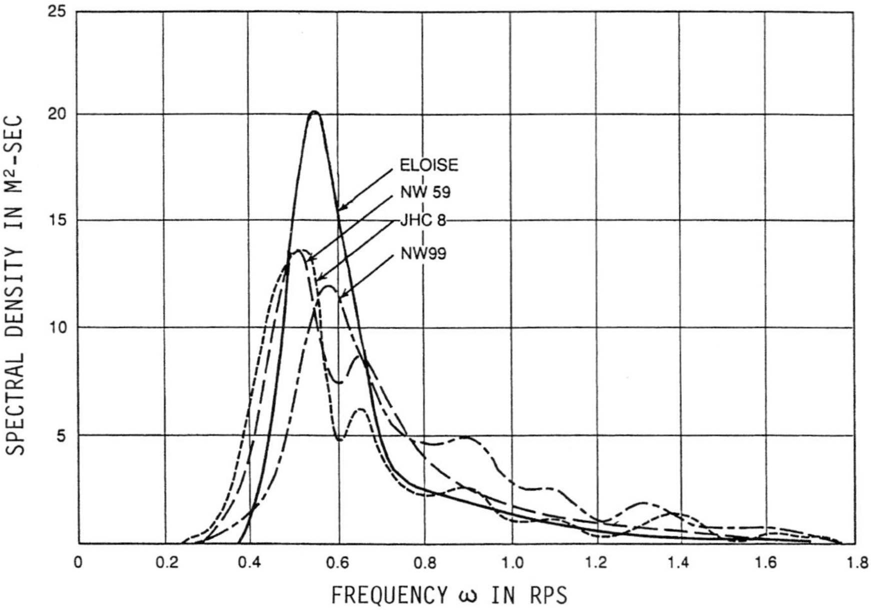

Figure 2.1 – Comparison between hurricane ELOISE wave spectrum and spectra in ordinary wind-generated seas having approximately the same significant wave height 8.8 m (Ochi, 1993).

energy (modal frequency) is essentially constant while the wave energy is propagating. Let us examine this interesting and important feature in detail in the following.

Figure 2.2 shows the variation in the wave spectrum during the approaching stage of hurricane ELOISE to Buoy EB-10. As seen in the figure, the modal frequency is almost constant, $f_m = 0.09$ Hz ($\omega_m = 0.57$ rps) in this example for 4 h in which the significant wave height increased rapidly from 5.08 to 8.78 m. The wind speed increased from 20.2 to 35.2 m/s during these 4 h. Such a consistent rapid increase in wind speed and sea severity cannot be observed in ordinary storm seas. Now, the group velocity of the wave component having a modal frequency of 0.09 Hz is 8.67 m/s (16.9 knots). On the other hand, hurricane ELOISE was advancing with a speed of 5.92 m/s (11.5 knots). Since the group velocity is greater than the advancing speed of the hurricane, the swell having a frequency of 0.09 Hz propagates ahead of the hurricane with substantial energy. This results in the energy of the wave spectrum measured in the approaching path of the hurricane being augmented at the modal frequency to a great extent.

Another example of this feature observed in hurricane GLORIA is shown in Figure 2.3. The data were obtained by Buoy 41002 (32.3N, 75.3W) when GLORIA was approaching. At this time GLORIA had a mean wind speed of 28.5 m/s (55.4 knots) which is below the normal wind speed of a hurricane, but the sea state was extremely severe as mentioned in Section 1.2. As seen in the figure, the shape of the wave spectra shows that energy is sharply concentrated in the neighborhood of the modal frequency. The significant wave height rapidly increases from 8.76 to 14.32 m in only 5 h, but the modal frequency maintains its constant value of 0.065 Hz in this case.

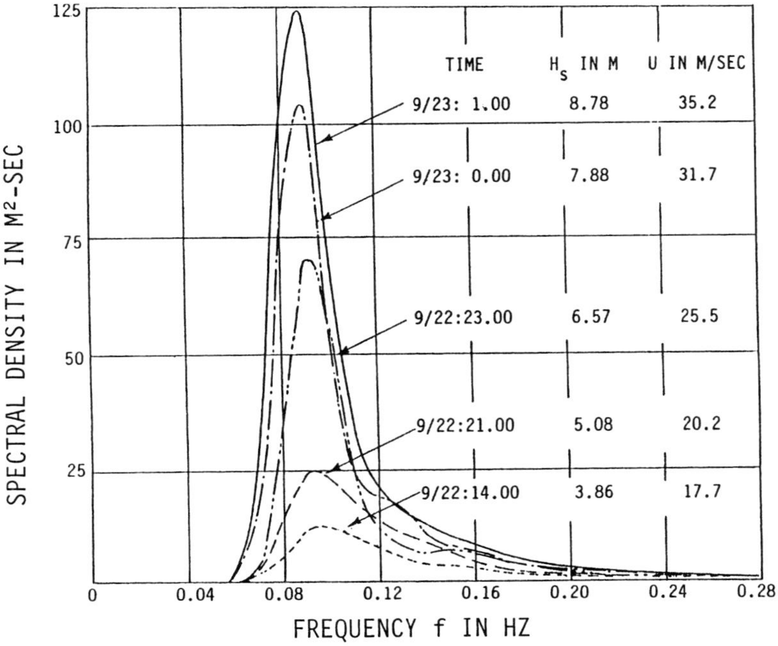

Figure 2.2 – Growing state of wave spectra measured at buoy EB10 during the last 10 h preceding the approach of hurricane ELOISE.

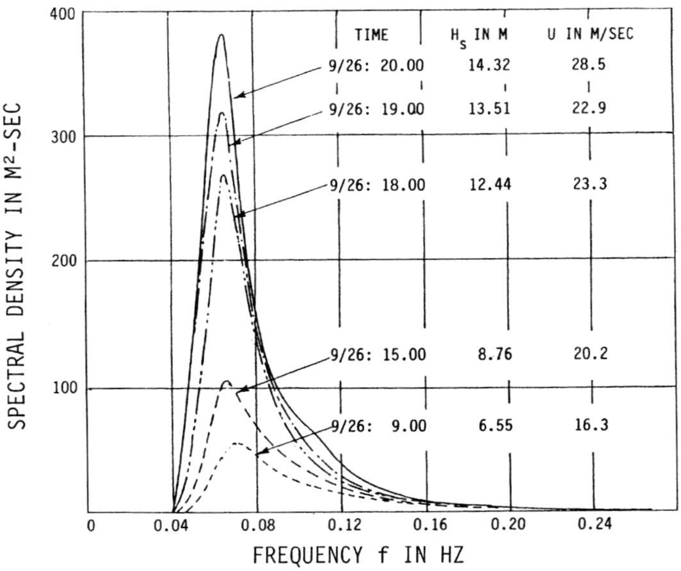

Figure 2.3 – Growing state of wave spectra measured at buoy 41002 during the last 11 h preceding the approach of hurricane GLORIA.

It is of interest to note that although the modal frequency in severe seas is constant for both ELOISE and GLORIA; there exists a significant difference between them; 0.09 Hz (0.57 rps) for ELOISE and 0.065 Hz (0.41 rps) for GLORIA. The results of analysis show that the value of the modal frequency in hurricane wave spectra depends on the magnitude of wind speed and sea severity. This subject will be discussed in Section 2.3.

The third feature of hurricane associated wave spectra is that, at the growing stage of the wave spectra, the increasing rate of wave energy in the neighborhood of the modal frequency is much greater than that at any other frequency of the spectrum. This feature can be seen in both wave spectra shown in Figures 2.2 and 2.3. A study on this subject was carried out in detail by Finlayson (1997) with the aid of a mathematical formulation of the hurricane-generated wave spectrum which will be presented in Section 2.2. Hence, it may be well to discuss this feature of hurricane wave spectra based on the Finlayson's study at the end of Section 2.2.

2.2 Presentation of wave spectrum by mathematical formulation

It is of importance to present the shape of wave spectra observed during a hurricane by a mathematical formulation so that the wave characteristics (including the maximum wave height) can be evaluated. Although many studies on forecasting and hindcasting wave spectra of hurricane-generated seas have been carried out, discussion in this section addresses only those studies on the mathematical presentation of spectra based on analysis of measured data.

First, it is noted that the results of all studies on hurricane-generated wave spectra obtained from analysis of measured data show that the wave spectra can best be represented in the form of the JONSWAP spectral formulation (Hasselmann *et al*, 1973). However, the values of the parameters involved in the JONSWAP spectra referred to here are not those given in the original publication; instead, the values are appropriately determined from hurricane data for each spectrum. Since the JONSWAP formulation plays an important role in representing hurricane-generated seas, it may be well to present the original formulation as follows.

$$S(f) = \alpha \frac{g^2}{(2\pi)^4} \frac{1}{f^5} \exp\{-1.25(f_m/f)^4\} \gamma^{\exp\{-(f-f_m)^2/2(\sigma f_m)^2\}} \tag{2.2-1}$$

where

γ	peak shape parameter, 3.30 as an average
f_m	modal frequency = 3.5 (g/U) $\bar{X}^{-0.33}$
α	$0.076\bar{X}^{-0.22}$
σ	0.07 for $f \le f_m$ and 0.09 for $f > f_m$
$\bar{X}$	dimensionless fetch = gX/U^2
U	mean wind speed
g	gravity constant

The formula may be expressed in terms of the frequency ω in radian per second (rps) as

$$S(\omega) = \alpha \frac{g^2}{\omega^5} \exp\{-1.25\{\omega_m/\omega)^4\} \gamma^{\exp\{-(\omega-\omega_m)^2/2(\sigma\omega_m)^2\}} \tag{2.2-2}$$

where

$$\omega_m = 2\pi f_m$$

An extensive analysis of hurricane-associated wave spectra was carried out by Antani (1981) in which he presents spectra computed from wave data obtained by NOAA buoys in hurricanes ELOISE, FREDERICK, ANITA, BABE and CAMILLE by applying a variety of spectral formulations. These include formulations in the form of Pierson–Moskowitz fully developed seas (Pierson and Moskowitz, 1964), the two-parameter spectrum (originally developed by Bretschneider (1959)), the six-parameter spectrum (Ochi and Hubble, 1976) and the JONSWAP spectrum (Hasselmann *et al*, 1973). In addition, Antani developed a five-parameter spectrum. From the results of analysis, Antani finds that the mathematical presentation in the form of six parameters and the JONSWAP formulation well represent a wide variety of spectral shapes over the full frequency range associated with stages of growth, eye passage and decay of hurricane-generated seas. In the following, analysis carried out by applying the JONSWAP spectral formulation is presented.

In representing a wave spectrum by the JONSWAP formulation without knowledge of wind speed and fetch length, the parameter α is expressed as a function of γ from the magnitude of spectrum at the modal frequency, $S(f_m)$, i.e. at the modal frequency, we have from Equation (2.2-1)

$$S(f_m) = \alpha \frac{g^2}{(2\pi)^4} \frac{1}{f_m^5} e^{-1.25} \gamma \tag{2.2-3}$$

Since $S(f_m)$ is known from data, we have

$$\alpha = S(f_m) \frac{(2\pi)^4 f_m^5}{g^2} e^{1.25} (1/\gamma) \tag{2.2-4}$$

Hence, we may write Equation (2.2-1) as

$$S(f) = S(f_m) \left(\frac{f_m}{f} \right)^5 e^{-1.25\{(f_m/f)^4 - 1\}} \gamma^{\exp\{-(f-f_m)^2/2(\sigma f_m)^2\} - 1} \tag{2.2-5}$$

Here, the unknown parameter γ can be determined for a given spectrum by letting its initial value be 0.8 with an increment of 0.05 over the range of $0.8 \le \gamma \le 6.0$. By letting the σ-values be those defined in the original paper of the JONSWAP spectrum, a spectrum can be computed for each γ-value. Then, we may determine the parameter γ which yields the best fit to the measured spectrum through the least-square technique. Examples of comparison between the measured spectrum and a representative JONSWAP-form spectrum computed by following the above method for hurricanes CAMILLE and ELOISE are shown in Figure 2.4.

Ross (1979), on the other hand, applies Equation (2.2-1) for representing hurricane-generated seas by evaluating the values of all parameters, α, γ, f_m, and σ in the formulation by the least square fitting technique. Examples of the fitting results are shown in Figure 2.5 in which the spectral density is normalized such that the density is unity at the peak frequency. As seen, the agreement between data and the JONSWAP formulation is excellent. From the results of regression analysis of thus computed parameters, Ross presents the scale parameter α and the peak shape parameter γ as a function of the dimensionless modal frequency $\bar{f}_m$ and the dimensionless distance $\bar{X}$, respectively, as follows

$$\alpha = 0.035 \bar{f}_m^{0.82} \tag{2.2-6}$$

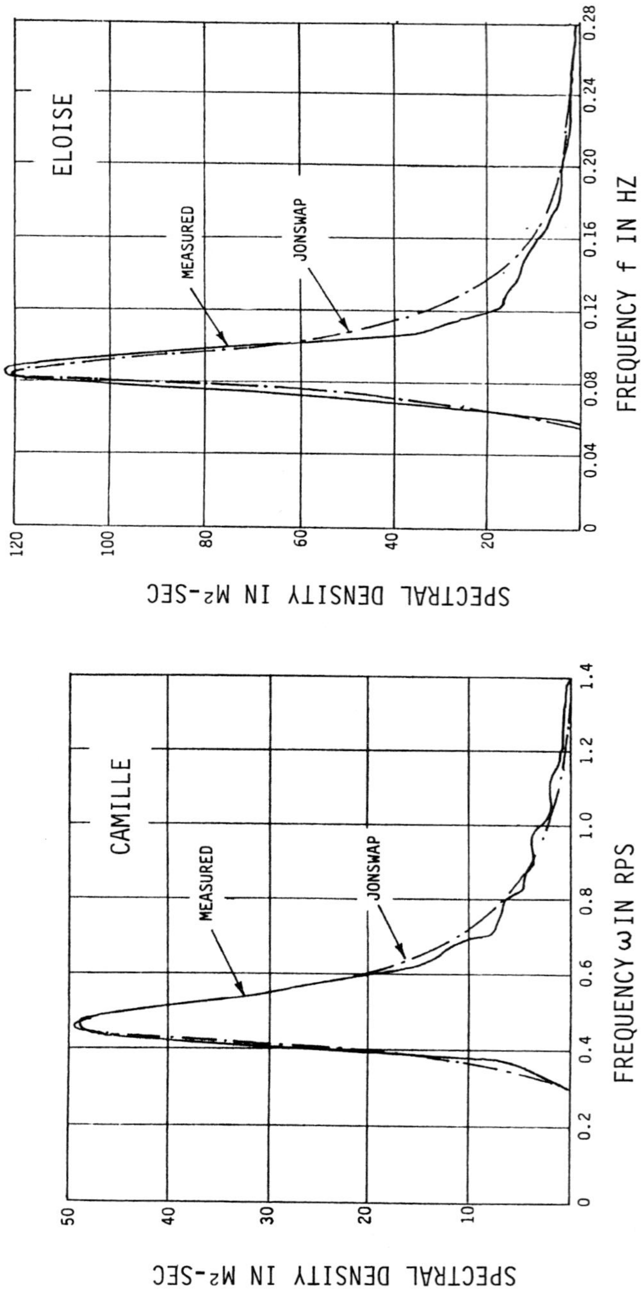

Figure 2.4 – Comparison between measured and computed JONSWAP-form spectra (Antani, 1981).

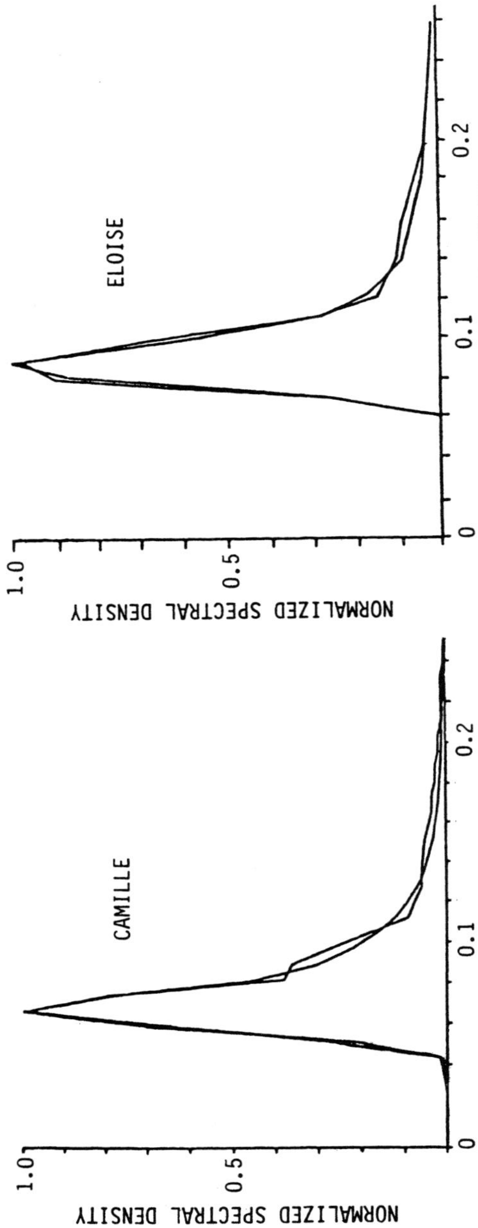

Figure 2.5 – Comparison between measured and computed JONSWAP-form spectra (Ross, 1979).

$$\gamma = 4.7\bar{X}^{-0.13}$$

where

$$\bar{f}_{\mathrm{m}} = f_{\mathrm{m}} U/g, \qquad \bar{X} = Xg/U^2$$

Further, the dimensionless modal frequency $\bar{f}_{\mathrm{m}}$ is found to be a function of the dimension-less distance $\bar{X}$ as shown in Figure 2.6, i.e.

$$\bar{f}_{\mathrm{m}} = 0.97\bar{X}^{-0.21} \tag{2.2-7}$$

Hence, the scale parameter becomes a function of the dimensionless distance expressed as

$$\alpha = 0.034\bar{X}^{-0.172} \tag{2.2-8}$$

The dimensionless modal frequency given in Equation (2.2-7) will be compared with that derived from analysis of some other data in Section 2.3.

In regard to the severity of sea state generated by hurricanes, Ross derives the following formula which is shown in Figure 2.7.

$$\bar{m}_0 = 2.5 \times 10^{-5}\bar{X}^{0.45} \tag{2.2-9}$$

where $\bar{m}_0 = m_0 g^2/U^4$, and m_0 is area under the spectrum.

Ross states that the scatter of data may be largely attributed to errors involved in estimating wind speed. The maximum error which may occur in the functional relationship resulting from a difference in the amount of ± 5 m/s is given in the figure. Although Equation (2.2-9) is the average value of the scattered data, the relationship of sea severity and the dimensionless fetch length reasonably agrees with that obtained from many other data as will be shown later.

Lee (1980) analyzes a limited amount of wave data for wind speeds over 30 knots (15.4 m/s) obtained by the NOAA buoy during hurricane ELOISE, and derives the dimensionless modal frequency as a function of the dimensionless fetch as

$$\bar{f}_{\mathrm{m}} = 1.15\bar{X}^{-0.22} \tag{2.2-10}$$

In his analysis, the least square fitting technique is employed. Furthermore, he presents the scale parameter α and the peak shape parameter γ of the JONSWAP formulation

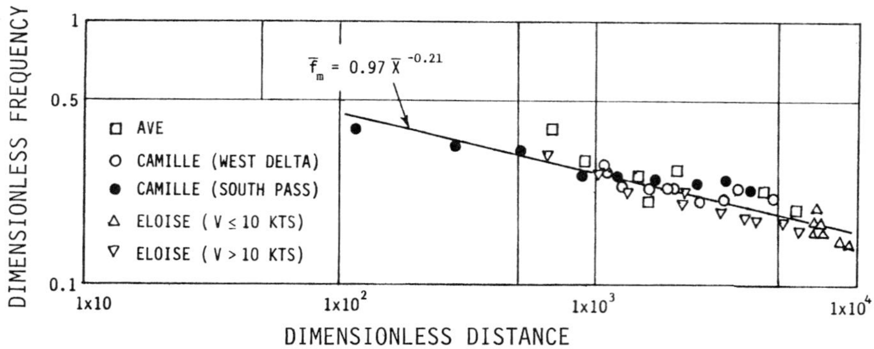

Figure 2.6 – Dimensionless modal frequency $\bar{f}_{\mathrm{m}}$ as a function of dimensionless distance $\bar{X}$ (Ross, 1979).

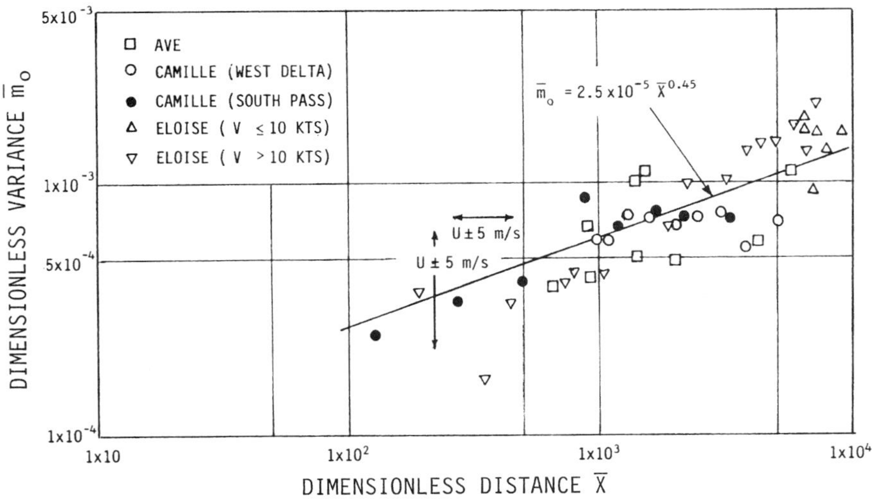

Figure 2.7 – Dimensionless variance $\bar{m}_0$ as a function of dimensionless distance $\bar{X}$ (Ross, 1979).

as follows

$$\alpha = 0.037\bar{X}^{-0.18} \qquad \gamma = 9.88\bar{X}^{-0.22} \tag{2.2-11}$$

It is noted here that f_m and α reasonably agree with that derived by Ross in the range of dimensionless distance (fetch) $\bar{X}$ from 2×10^2 to 4×10^4. However, the γ-values are quite different for small and large values of the dimensionless distance.

Young (1998) compiles a large number of spectra obtained primarily in tropical cyclones off the north-west coast of Australia. The JONSWAP formulation is also considered in Young's analysis, but the parameter α and γ are analyzed as a function of U/c_p (the inverse of wave age), where c_p is the phase speed of the component wave at the modal frequency of the wave spectrum. He compares α and γ values evaluated from measured data with the following Donelan, Hamilton, and Hui formulation (1985)

$$\alpha = 0.006(U/c_p)^{0.55} \tag{2.2-12}$$

$$\gamma = \begin{cases} 1.7 & \text{for } 0.83 \le U/c \le 1 \\ 1.7 + 6\log_{10}(U/c_p) & \text{for } 1 \le U/c \le 5 \end{cases} \tag{2.2-13}$$

Although the α and γ values evaluated from measured data scatter, the average values reasonably agree with the formula given by Donelan *et al.* It is noted that for deep water the α-value given in Equation (2.2-12) can be expressed in terms of the dimensionless modal frequency as follows

$$\alpha = 0.006(U/c_p)^{0.55} = 0.006\left(\frac{2\pi f_m U}{g}\right)^{0.55} = 0.0165\bar{f}_m^{0.55} \tag{2.2-14}$$

The α-value thus presented, however, differs significantly from that of the formula derived by Ross and the γ-value computed by Donelan's formula is much larger than that computed by others. It is not clear at this stage if the difference may be attributed to the difference in sea severities associated with tropical cyclones which have much milder sea severity than hurricanes.

Foster (1982) carries out an exhaustive analysis of the parameters involved in the JONSWAP spectral formulation as applied to hurricane-generated seas and derives several important conclusions. One of his most important findings lies in the relationship between the scale parameter α and significant wave height H_s. In the analysis, Foster segregates α into six groups of modal frequency ranging from 0.07 Hz (0.44 rps) to 0.16 Hz (1.00 rps), and obtains the relationship for each group. As an example, Figure 2.8 shows the relationship between α and H_s obtained for $f_m = 0.08$ Hz (0.50 rps) and 0.13 Hz (0.82 rps). As seen in this figure, the scale parameter α and the significant wave height, H_s have a clear functional relationship for a given modal frequency.

Figure 2.9 shows the relationship between α and H_s plotted on log–log paper for various modal frequencies f_m. As seen, the entire set of lines can be reasonably considered to have a constant slope of 2.0 as its value. Thus, the scale parameter α of the JONSWAP formulation suitable for hurricane-associated wave spectra can be presented as

$$\alpha = c(f_m)H_s^2 \tag{2.2-15}$$

where $c(f_m) = 4.5 f_m^4$ as shown in Figure 2.10.

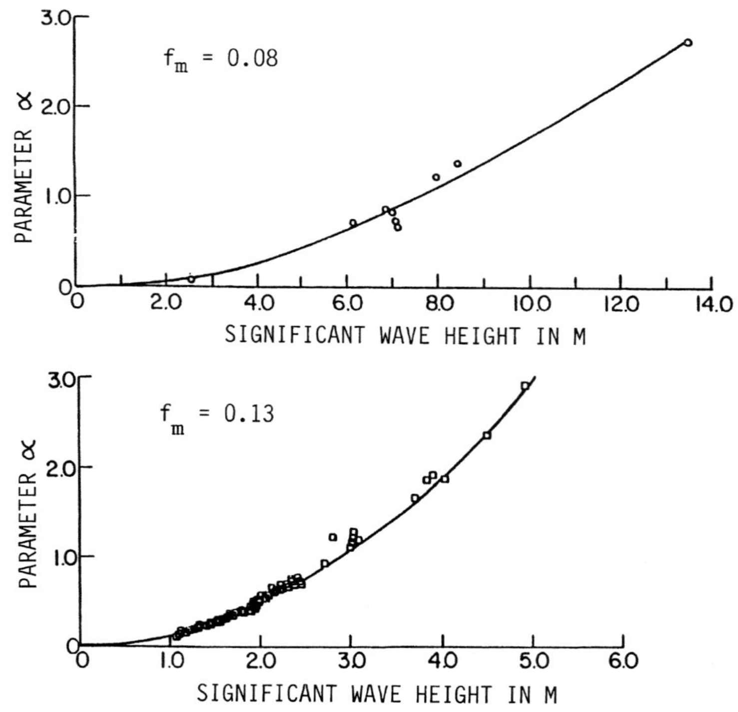

Figure 2.8 – Example of JONSWAP spectral scale parameter α as a function of significant wave height for a given modal frequency (Foster, 1982).

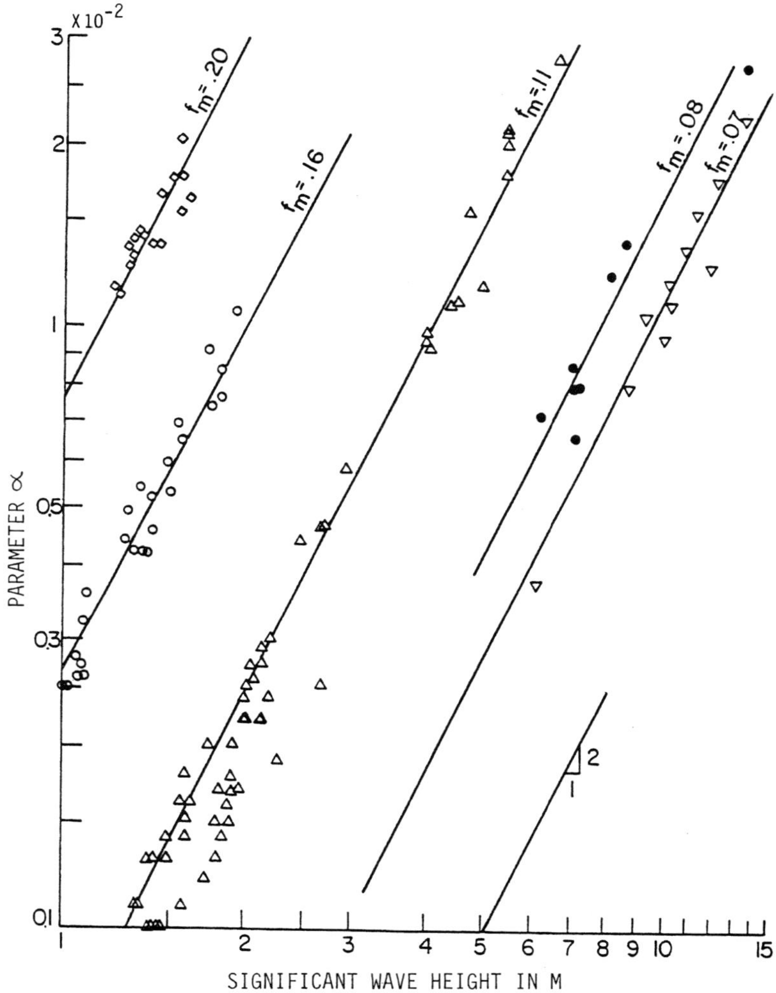

Figure 2.9 – JONSWAP spectral parameter α as a function of significant wave height for various modal frequency of hurricane-generated seas (Foster, 1982).

Hence, in summary, the parameter α may be expressed as

$$\alpha = 4.5 f_m^4 H_s^2 \tag{2.2-16}$$

It is noted that the constant involved in Equation (2.2-16) carries the unit of $(s^2/m)^2$.

In non-dimensional form, Equation (2.2-16) may be written as

$$\alpha = 432.2 \bar{f}_m^4 \bar{H}_s^2 \tag{2.2-17}$$

where $\bar{H}_s = H_s g/U^2$, $\bar{f}_m = f_m U/g$, and U is mean wind speed at 10 m height.

The peak shape parameter γ of the JONSWAP formulation as applied to hurricane-generated seas varies considerably depending on the sea severity. A statistical analysis of

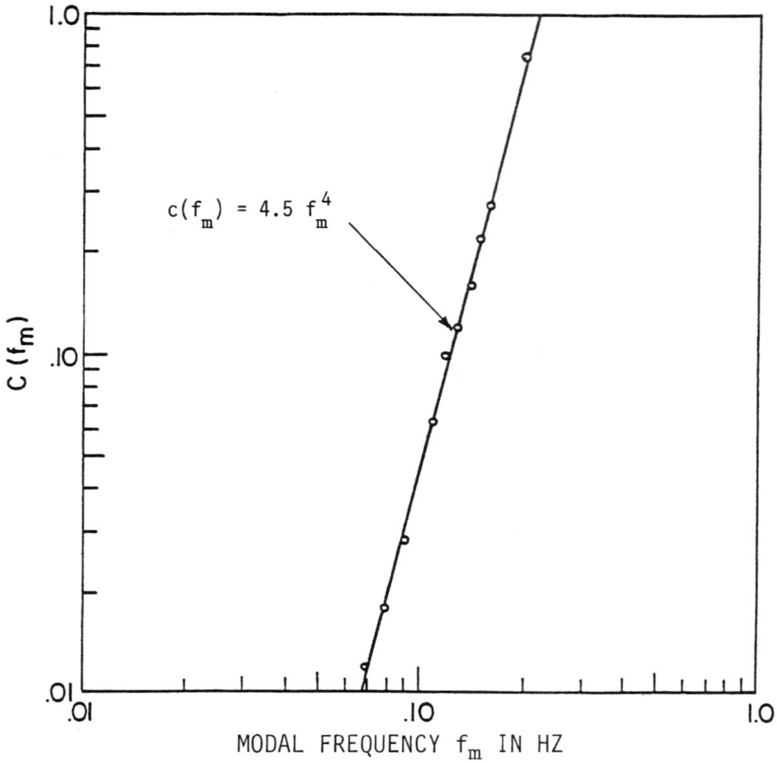

Figure 2.10 – Parameter for evaluating α as a function of modal frequency (Foster, 1982).

the γ-value is carried out by Foster, and its histogram is shown in Figure 2.11. As seen, the value of γ applicable to hurricane-generated seas ranges from 0.6 to 4.0; the range being significantly smaller than that obtained in the JONSWAP Project. For comparison purposes, the histogram of the parameter γ constructed from analysis of the original JONSWAP data is shown in Figure 2.12 (Ochi, 1979). As seen in the figure, the γ-value of the original JONSWAP data ranges from 1.0 to 6.0 with an average value of 3.30; but Figure 2.11 shows that the probability of occurrence of a γ-value in the neighborhood of 3.3 is extremely small in hurricane associated seas.

Doong *et al* (2001) analyze wave spectra obtained from measured data during three typhoons (hurricanes) in the pacific off the coast of Taiwan. By representing the spectra by the JONSWAP formulation, they find that the γ-value ranges from 0.5 to 2.8 with an average of 1.25 which agrees very well with the data shown in Figure 2.11.

In order to present the parameter γ as a function of hurricane-generated sea severity, Foster first derives the functional relationship between the peak energy density $S(f_m)$ and significant wave height H_s, where H_s is chosen to be greater than 2.5 m in order to eliminate the possibility of having double peaks in the spectra. It is found that there is a strong correlation between the peak density and significant wave height irrespective of the value of the modal frequency as shown in Figure 2.13. Foster presents the functional relationship of

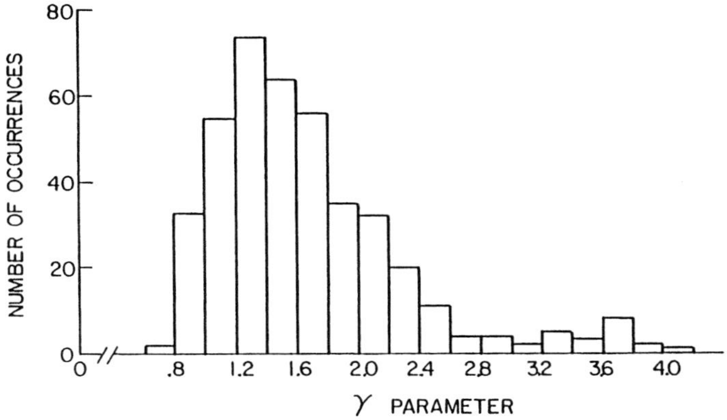

Figure 2.11 – Histogram of peak shape parameter γ of the JONSWAP formulation as applied to hurricane-generated seas (Foster, 1982).

the data shown in the figure as

$$S(f_m) = 0.756H_s^{2.34} \tag{2.2-18}$$

On the other hand, the peak energy density of the JONSWAP formulation is given in Equation (2.2-3). Hence, from Equation (2.2-18) along with Equations (2.2-3) and (2.2-16), Foster derives the parameter γ as a function of sea severity as follows

$$\gamma = 9.5f_m H_s^{0.34} \tag{2.2-19}$$

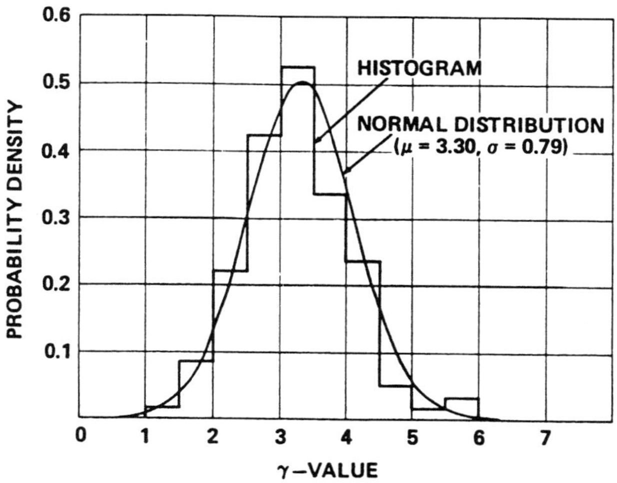

Figure 2.12 – Histogram of peak-shape parameter γ of the original JONSWAP data (Ochi, 1979).

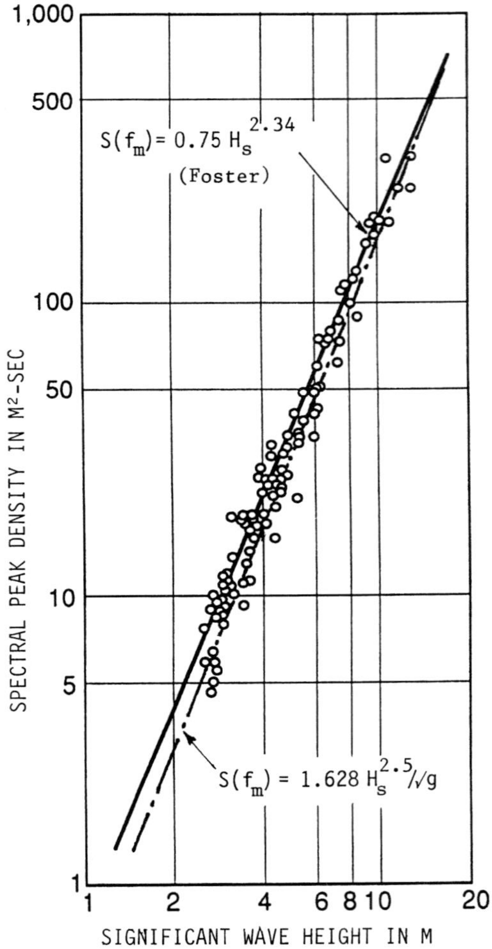

Figure 2.13 – Peak density of hurricane-generated wave spectra as a function of significant wave height (Foster, 1982).

It is noted that the significant wave height unit in Equation (2.2-19) should be in meters and that the constant in the equation carries a dimension. Unfortunately, it is not possible to express Equation (2.2-19) in dimensionless form of $\bar{f}_m$ and $\bar{H}_s$.

One way to present γ in dimensionless form is to express the peak value of the spectrum $S(f_m)$ given in Figure 2.13 approximately as

$$S(f_m) = 0.52 H_s^{2.50} = 1.628 H_s^{2.50} / \sqrt{g} \qquad (2.2\text{-}20)$$

The above equation is also included in Figure 2.13. Then, from Equations (2.2-3), (2.2-17), and (2.2-20), the following formula can be derived.

$$\gamma = 20.48 \bar{f}_m \sqrt{\bar{H}_s} \qquad (2.2\text{-}21)$$

In this case, γ can be written in dimensional form, f_m and H_s, as follows

$$\gamma = 6.54 f_m \sqrt{H_s} \tag{2.2-22}$$

The difference in the γ-value computed by Equations (2.2-19) and (2.2-22) is very small. For the severest sea state observed in hurricane GLORIA ($H_s = 14.32$ m, shown in Figure 2.3) the difference in the γ-value is 5.1%, and for hurricane ELOISE ($H_s = 8.78$ m, shown in Figure 2.2) the difference is 2.3%. Thus, we may consider Equation (2.2-19) for further development of a formula applicable to hurricane-generated wave spectra, although the constant carries a dimension.

By applying the parameters α and γ given in Equations (2.2-16) and (2.2-19), respectively, wave spectra applicable for hurricane-generated seas can be presented in the form of the JONSWAP formulation as a function of significant wave height and modal frequency. This may be called the modified JONSWAP spectral formulation and is given as follows (Ochi, 1993)

$$S(f) = \frac{4.5}{(2\pi)^4} g^2 H_s^2 \frac{f_m^4}{f^5} e^{-1.25(f_m/f)^4} (9.5 f_m H_s^{0.34})^{\exp\{-(f-f_m)^2/2(\sigma f_m)^2\}} \tag{2.2-23}$$

where the units of H_s and f_m are meters and Hz, respectively. Furthermore, the formula can be written in terms of frequency ω (rps) as follows

$$S(\omega) = \frac{4.5}{(2\pi)^4} g^2 H_s^2 \frac{\omega_m^4}{\omega^5} e^{-1.25(\omega_m/\omega)^4} \left(\frac{9.5}{2\pi} \omega_m H_s^{0.34}\right)^{\exp\{-(\omega-\omega_m)^2/2(\sigma\omega_m)^2\}} \tag{2.2-24}$$

Examples of comparisons between wave spectral formulation for hurricane-generated seas (modified JONSWAP) and spectra obtained from measured data in the severest sea during hurricanes ELOISE, KATE, GLORIA and GEORGE are shown in Figures 2.14–2.17, respectively. As seen in these figures, the spectral formulation represents measured data very well.

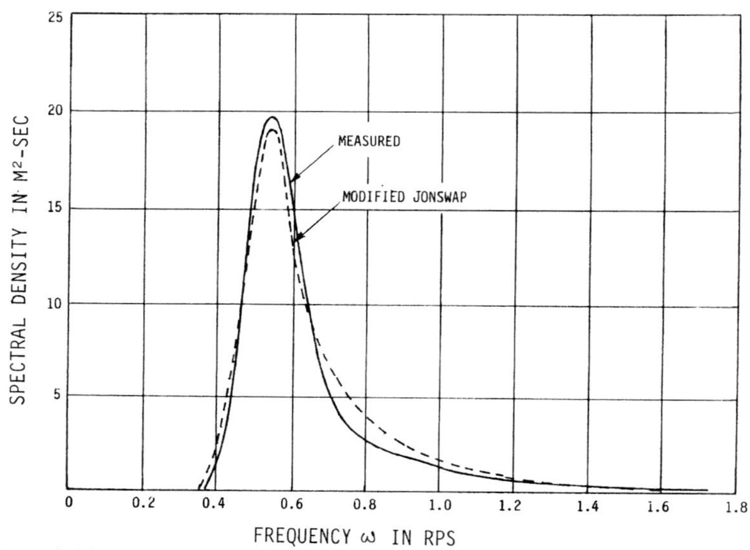

Figure 2.14 – Comparison between spectral formulation for hurricane-generated seas and measured spectrum during hurricane ELOISE.

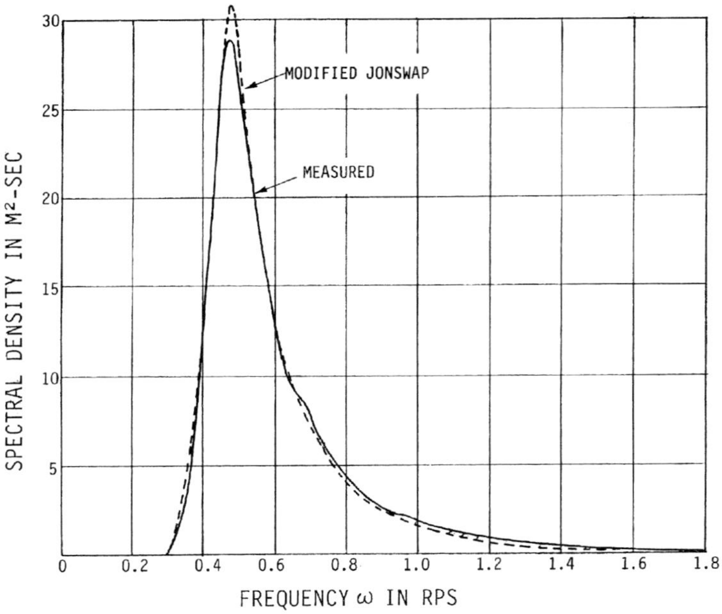

Figure 2.15 – Comparison between spectral formulation for hurricane-generated seas and measured spectrum during hurricane KATE.

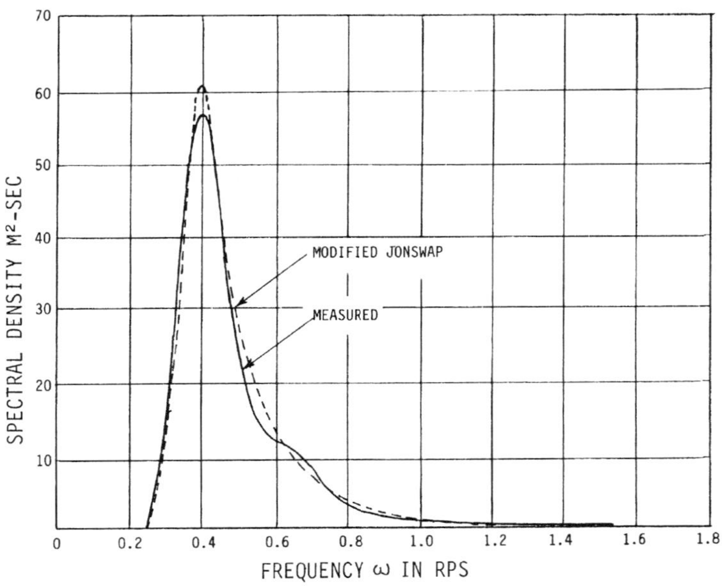

Figure 2.16 – Comparison between spectral formulation for hurricane-generated seas and measured spectrum during hurricane GLORIA.

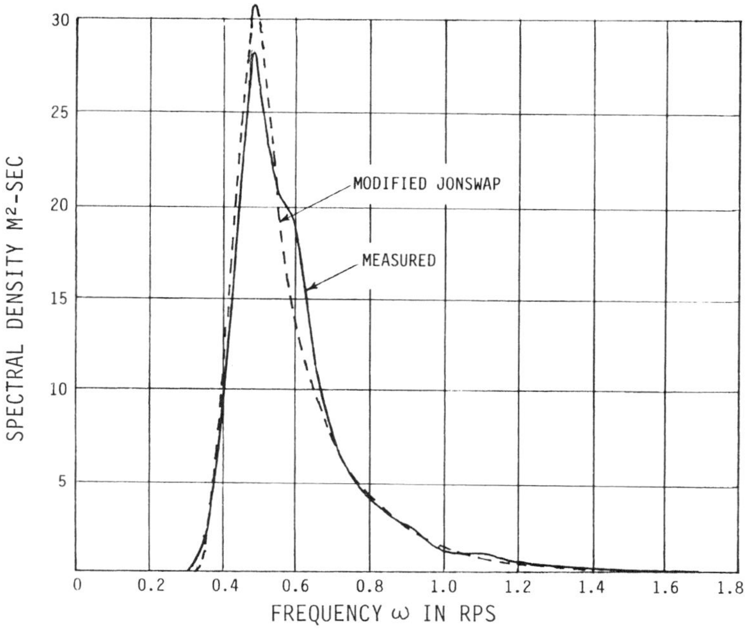

Figure 2.17 – Comparison between spectral formulation for hurricane-generated seas and measured spectrum during hurricane GEORGE.

A study on the comparison of spectra obtained from measured data and the modified JONSWAP spectral formulation is carried out by Finlayson (1997) in great detail. The focus of his analysis lies in examining how well the spectral formulation represents measured data not only in the neighborhood of the modal frequency but also over the entire range of the frequency domain. For this, the frequency range of each spectrum obtained from measured data is normalized into 30 sections with an interval of 0.05 rps and presented in a dimensionless ratio ω/ω_m, where ω_m is the modal frequency. The spectral density obtained from measured data for each frequency ratio is plotted as a function of the significant wave height, and the results are compared with the spectral density function computed by the formulation.

Figure 2.18 taken from Finlayson's publication shows examples for six values of ω/ω_m. As seen in these examples, the spectral formulation agrees reasonably well with measured data throughout the normalized frequency domain. At frequency ratios higher than 2.0 the data begins to scatter. However, it is recognized that the magnitude of the spectral density is small at these frequencies; hence, the scatter of data and disagreement with the computed values should not cause a serious problem. Based on the results of Finlayson's study, it is concluded that the spectral formulation given in Equations (2.2-23) and (2.2-24) well represent hurricane-generated seas in deep water over almost the entire frequency range.

The shape of wave spectra associated with hurricane-generated seas in deep water usually has a single peak except in mild sea conditions (say, significant wave height less than 2.5 m) at an early stage of a hurricane. However, the shape of wave spectra in finite

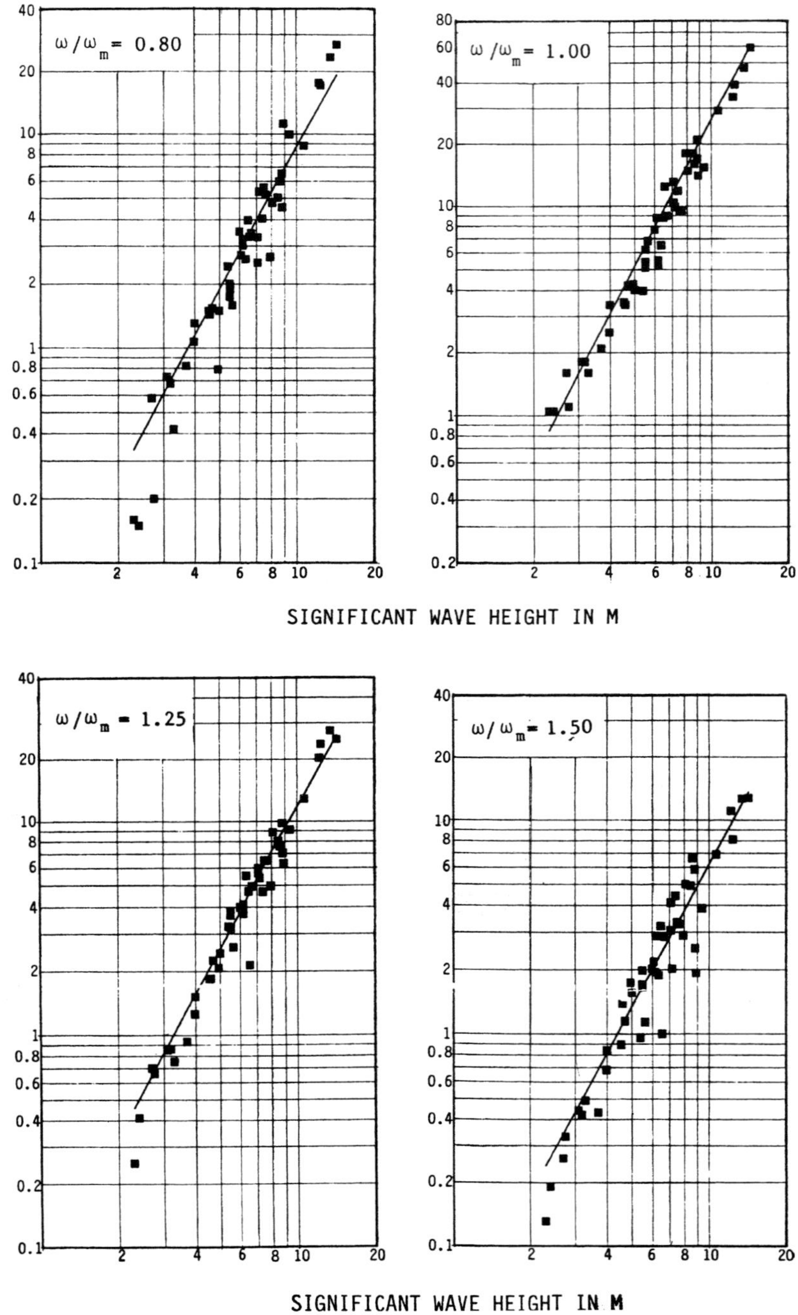

Figure 2.18 – Comparison between computed spectral density as a function of significant wave height (solid line) and measured data for $\omega/\omega_m = 0.80$, *1.00, 1.25, 1.50, 2.00, and 2.25 (Finlayson, 1997).*

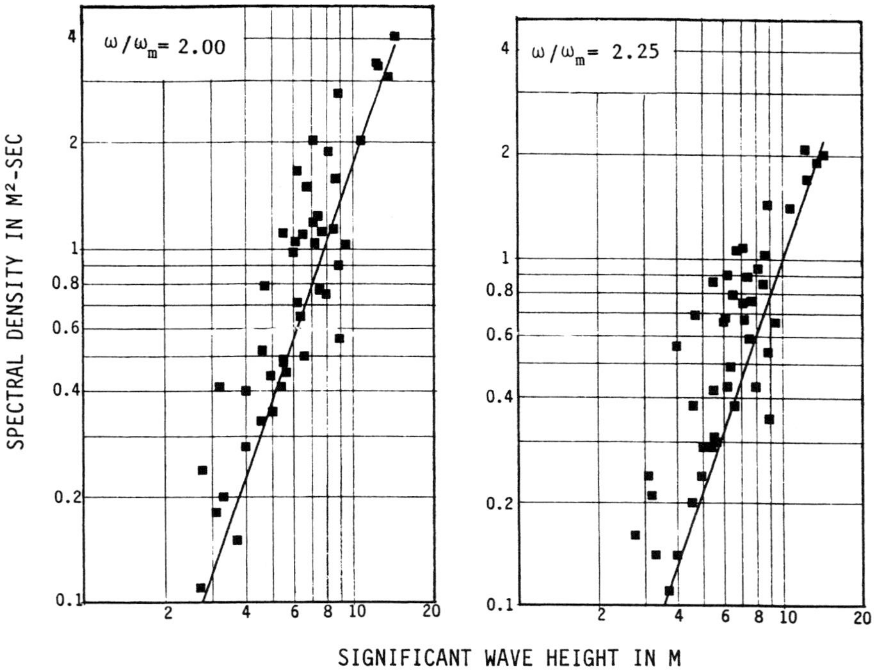

Figure 2.18 (continued).

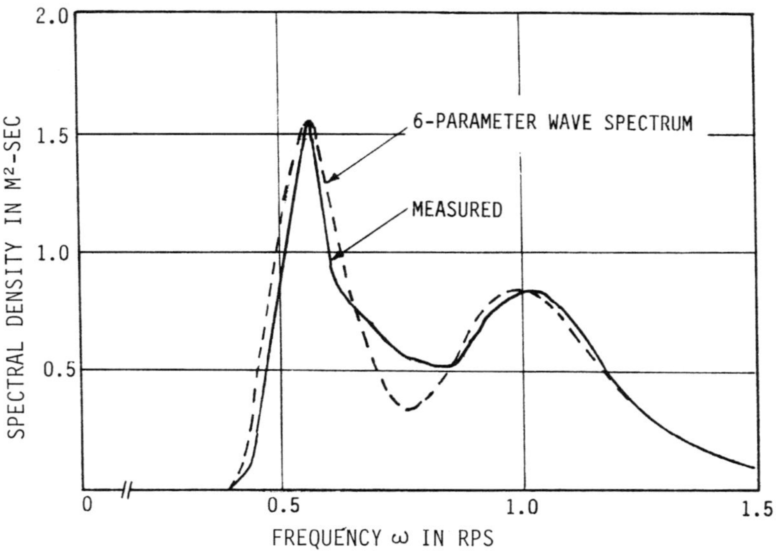

Figure 2.19 – Comparison between double-peaked spectrum measured during hurricane EDITH and six-parameter wave spectral formulation (Whalen and Ochi, 1978).

water depth, particularly in the nearshore area, sometimes shows double peaks since the local sea and swell traveling ahead of the hurricane are mixed. These double-peaked spectra are represented reasonably well by applying the six-parameter wave spectral formulation. As an example the double-peaked spectrum obtained at Station 6 in the Ocean Data Gathering Program during the early stage of hurricane EDITH is shown in Figure 2.19 (Whalen and Ochi, 1978).

2.3 Wave spectra for design consideration of marine systems

Estimation of sea state including the extreme sea condition generated by hurricane winds is a challenging topic in naval, ocean and coastal engineering, and in the past several decades many researchers have devoted their effort to provide information vital for the design of marine systems and coastal structures. Among others are Bretschneider (1972), Bea (1974), Ward, Borgman, and Cardon (1979), and Young (1988a,b). We now have more measured data and improved knowledge on hurricane wind spectra over the ocean as well as on wave spectra representing severe sea conditions generated by hurricanes. In the following, a method to estimate wave spectra for design consideration of marine systems is presented based on recently available techniques.

From the discussion presented in Section 2.2, it is clear that hurricane-generated wave spectra in deep water are well represented by the modified JONSWAP spectral formulation as demonstrated in Figs. 2.14–2.17. The values of parameters involved in the formulation are substantially different from those originally specified in the JONSWAP formulation which is applicable for ordinary wind-generated seas. The modified JONSWAP spectral formulation are presented in terms of significant wave height, H_s, and modal frequency, f_m, at a specified location in deep water. In order to apply the formulation for the design of marine systems, it is necessary to present the significant wave height and modal frequency as a function of hurricane-associated parameters such as wind speed, distance from the site to the hurricane center (fetch length), etc.

Le us first find the relationship between the modal frequency and the fetch length. Figure 2.20 shows the results of analysis of data obtained during the growing stage of hurricanes KATE, ELOISE, FREDERICK and GLORIA. In the figure, the dimensionless modal frequency $\bar{f}_m = f_m U/g$ is plotted against the dimensionless fetch length $\bar{X} = Xg/U^2$. As mentioned in detail in Section 1.2, the sea state before hurricanes KATE, ELOISE and FREDERICK arrived at the measurement site was relatively calm, while the sea prior to the arrival of hurricane GLORIA was not calm; instead, a sea of significant wave height on the order of 2–3 m lasted for more than 10 days. As seen in Figure 2.20, there is a distinct difference in the functional relationships depending on the sea condition prior to hurricane arrival at the site.

As shown earlier in Figure 2.6, Ross (1979) gives the following functional relationship between the dimensionless modal frequency and fetch length based on his analysis of hurricanes AVE, CAMILLE and ELOISE,

$$\bar{f}_m = 0.97\bar{X}^{-0.21} \tag{2.3-1}$$

The above equation is also given in Figure 2.20. As can be seen, Ross's formula represents well many other data which are not included in his analysis. These data were all obtained at locations where the sea condition prior to hurricane arrival was relatively calm for at least one week.

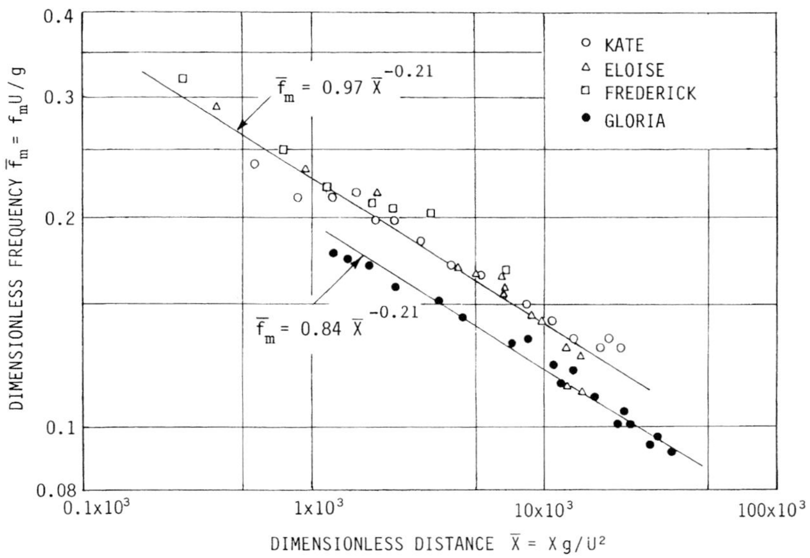

Figure 2.20 – Functional relationship between dimensionless modal frequency and dimensionless distance.

For data obtained at a location where the prior sea condition was not calm and a significant wave height of 2–3 m continued for more than 10 days, we may write the functional relationship between dimensionless modal frequency and fetch length in a form similar to Equation (2.3-1) as follows

$$\bar{f}_m = 0.84\bar{X}^{-0.21} \tag{2.3-2}$$

The relationships given in Equations (2.3-1) and (2.3-2) may be written in dimensional form as follows

$$f_m = \begin{cases} 5.89U^{-0.58}X^{-0.21} & \text{for relatively calm sea prior to hurricane arrival} \\ 5.10U^{-0.58}X^{-0.21} & \text{for moderate sea prior to hurricane arrival} \end{cases} \tag{2.3-3}$$

where f_m is in Hz, U in m/s, and X in meters.

Thus, the modal frequency of the wave spectrum at a specified location in deep water can be estimated from knowledge of the wind speed at the location and the distance to the hurricane center.

It may be of interest to examine how the modal frequencies of hurricane-associated wave spectra differ from those of ordinary wind-generated wave spectra. The modal frequency of the JONSWAP spectrum is given from analysis of fetch limited storm wave spectra as follows

$$f_m = 3.5(g^2/UX)^{-0.33} = 3.5(g/U)\bar{X}^{-0.33} \tag{2.3-4}$$

This can be written in dimensionless form of the frequency as

$$\bar{f}_m = 3.5\bar{X}^{-0.33} \tag{2.3-5}$$

which yields a much greater $\bar{f}_m$ than that given in Equation (2.3-1) for a specified dimensionless fetch length $\bar{X}$. For example, the value of the dimensionless frequency $\bar{f}_m$ is 0.35 for $\bar{X} = 1 \times 10^3$ as compared with 0.227 for a hurricane-associated sea computed by Equation (2.3-1). This implies that for the same wind speed and fetch length, the modal frequency of a hurricane-generated wave spectrum is much less, hence the length of waves having the maximum energy is much greater than that expected in ordinary storm seas.

Results of analysis carried out on the functional relationship between sea severity (significant wave height) and fetch length from data obtained during four hurricanes are shown in Figure 2.21. A dimensionless presentation of significant wave height $\bar{H}_s = H_s g/U^2$ is employed in the figure. Again, there is a distinct difference in the relationship depending on the sea condition prior to hurricane arrival at the site. The relationship between significant wave height and fetch length may be presented in dimensionless form as

$$\bar{H}_s = \begin{cases} 0.033\bar{X}^{0.162} & \text{for relatively calm sea prior to hurricane arrival} \\ 0.049\bar{X}^{0.162} & \text{for moderate sea prior to hurricane arrival} \end{cases} \tag{2.3-6}$$

In dimensional form, the relationships are written as

$$H_s = \begin{cases} 0.0048U^{1.68}X^{0.162} & \text{for relatively calm sea prior to hurricane arrival} \\ 0.0072U^{1.68}X^{0.162} & \text{for moderate sea prior to hurricane arrival} \end{cases} \tag{2.3-7}$$

where H_s and X are in meters and U is in m/s.

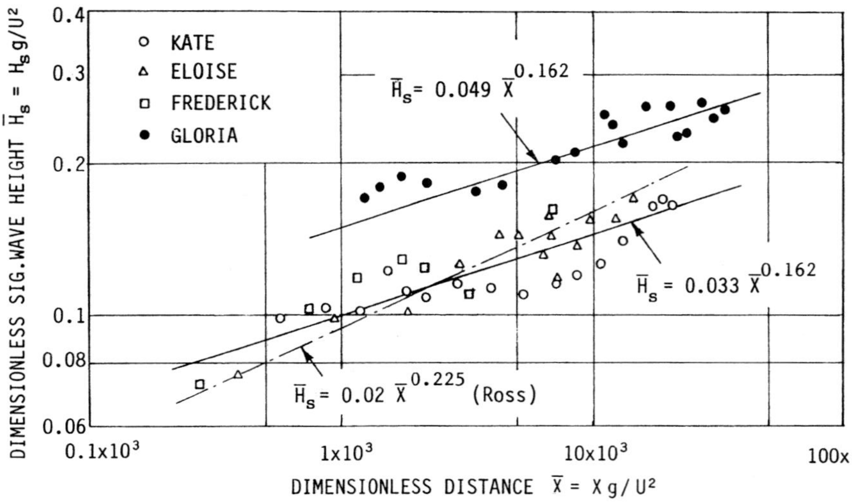

Figure 2.21 – Functional relationship between dimensionless significant wave height and dimensionless distance.

Ross (1979) derives the following functional relationship between sea severity and fetch length in dimensionless form,

$$\varepsilon = \frac{2.5}{10^5}\bar{X}^{0.45} \qquad (2.3\text{-}8)$$

where $\varepsilon = Eg^2/U^4$, E is the area under the spectral density function, and U the wind speed at 10 m height.

Assuming that the significant wave height can be presented as $H_s = 4\sqrt{E}$, Equation (2.3-8) can be written in terms of dimensionless significant wave height as

$$\bar{H}_s = \frac{H_s g}{U^2} = 0.02\bar{X}^{0.225} \qquad (2.3\text{-}9)$$

The relationship given in Equation (2.3-9) is included in Figure 2.21. As seen, Ross's formula also represents the data reasonably well. However, the formulae given in Equations (2.3-6) and (2.3-7) are preferable in that hurricane-generated sea severity is presented in the same functional form of wind speed and fetch length irrespective of sea condition prior to hurricane arrival.

Thus, with the aid of Equations (2.3-3) and (2.3-7), hurricane-generated wave spectra in deep water can be obtained by Equation (2.2-23) or (2.2-24) as a function of wind speed at the site and distance to the hurricane center.

Next, let us estimate the wind speed at the site. We may consider the simple case that the site is on the path of the approaching hurricane. Many studies have been carried out for estimating the magnitude of gradient wind speeds of a hurricane including several empirical formulae developed based on observed data; Graham and Nunn (1959), Kraft (1961), Collins and Viehman (1971), Atkinson and Holliday (1977), and Schwerdt, Ho, and Watkins (1979) among others. The wind field, in general, is estimated based on the pressure gradient and accurate estimation is extremely complicated. In the following, an approximate method for estimating hurricane wind profiles as a solution of the governing differential equation is outlined.

From the results of analysis of observed data, Schloemer (1954) presents normalized pressure profiles associated with hurricane winds as follows

$$\frac{p - p_c}{p_n - p_c} = \exp\{-R/r\} \qquad (2.3\text{-}10)$$

where p is pressure at radius r in mb, p_c the central pressure in mb, p_n the ambient (neutral) pressure in mb, R the radius of maximum wind speed, and r the distance from the center (radius).

In Equation (2.3-10), Wang (1978) defines R as one-tenth of the radial distance from the center to the periphery of the hurricane. Based on the results of his analysis he states that over the open ocean this distance is found to be near the 1000 mb isobar.

On the other hand, Holland (1980) employs a more general form in representing the normalized pressure as

$$\frac{p - p_c}{p_n - p_c} = \exp\{-A/r^B\} \qquad (2.3\text{-}11)$$

where A and B are constants to be determined from observed data. From the above equation, the derivative of p with respect to r yields,

$$\frac{dp}{dr} = (p_n - p_c)\left(\frac{AB}{r^{B+1}}\right)\exp\{-A/r^B\} \qquad (2.3\text{-}12)$$

Then, the wind velocity associated with hurricane denoted by V_r can be evaluated, by neglecting the surface friction, as a function of distance r by

$$V_r = \left(\frac{r}{\rho}\frac{\mathrm{d}p}{\mathrm{d}r}\right)^{1/2} = \left[\frac{1}{\rho}(p_\mathrm{n} - p_\mathrm{c})\frac{AB}{r^B}\exp\{-A/B^r\}\right]^{1/2} \tag{2.3-13}$$

where ρ is the air density (1.15 kg/m^3), and V_r is called the gradient wind velocity at radius r.

For a more accurate evaluation of the gradient wind velocity, the following geostrophic wind associated with the earth rotation, denoted by V_g, should be included.

$$V_\mathrm{g} = \frac{1}{\rho f}\frac{\mathrm{d}p}{\mathrm{d}r} = \frac{1}{\rho f}(p_\mathrm{n} - p_\mathrm{c})\frac{AB}{r^{B+1}}\exp\{-A/r^B\} \tag{2.3-14}$$

where f, the coriolis parameter $= 2\omega \sin\phi$, ω, the angular velocity of the earth $= 7.29 \times 10^{-5}$ rps and ϕ is the latitude.

The relationship between V_r and V_g is given in the northern hemisphere by

$$V_r^2 + rfV_r - rfV_\mathrm{g} = 0 \tag{2.3-15}$$

The gradient wind velocity V_r can then be obtained as a solution of Equation (2.3-15), i.e.

$$V_r = \left[\frac{1}{\rho}(p_\mathrm{n} - p_\mathrm{c})\frac{AB}{r^B}\exp\{-A/r^B\} + \left(\frac{rf}{2}\right)^2\right]^{1/2} - \frac{rf}{2} \tag{2.3-16}$$

The above equation is the gradient wind velocity at a distance r from the hurricane center developed by Holland. Further, he derives the radius of the maximum wind speed and the magnitude of the maximum wind. By neglecting the Coriolis force in the region of the maximum winds and by letting $\mathrm{d}V_r/\mathrm{d}r = 0$ in Equation (2.3-13), the radius of the maximum wind can be obtained as

$$r_\mathrm{max} = A^{1/B} \tag{2.3-17}$$

Then, from Equations (2.3-16) and (2.3-17), the magnitude of the maximum wind speed can be evaluated by

$$V_\mathrm{max} = \left[\frac{B}{\rho\,\mathrm{e}}(p_\mathrm{n} - p_\mathrm{c})\right]^{1/2} \tag{2.3-18}$$

where e is the base of the natural log, i.e. 2.718.

It is noted that the wind velocity given in Equations (2.3-16) and (2.3-18) is not the mean wind speed to be used for evaluating wind-generated wave spectra. To obtain the mean wind speed at 10 m height from the gradient wind velocity, Powell (1980) suggests the former may be estimated by multiplying the latter by a factor of 0.80. Another way for estimating the mean wind speed is to apply a gust factor to the gradient velocity, namely, by dividing the gradient velocity by the gust factor. This approach is considered here.

It is clear that reliable data on pressure gradients and accurate fitting of the normalized pressure by Equation (2.3-11) are required in order to apply Holland's method for estimating the gradient wind velocity. We may apply the method, however, for an approximate estimation of sea conditions by using the pressure data obtained by NOAA Buoy EB-10 during hurricane ELOISE in the following.

Figure 2.22 shows the normalized pressure profile data computed using the pressure data recorded at hourly intervals as a function of the estimated distance from the buoy to the hurricane center. In the computations, the central pressure and ambient pressure are assumed to be 967.8 and 1010 mb, respectively, as obtained from data. Unfortunately, the accuracy of the estimated distance is somewhat unreliable when the hurricane center was close to the buoy; say, less than 35 km. Included also in the figure is the functional presentation given in Equation (2.3-11) with parameters A and B determined from the data. As seen in the figure, it is not possible to present the entire data by a single formulation for this example. Hence, the data points for relatively short distances (less than 60 km) are presented by Equation (2.3-11) using different values of A and B determined from data. The gradient wind velocities are computed by Equation (2.3-16), and then the average wind velocities are obtained by using a gust factor of 1.23 obtained from the results of analysis of hurricane KATE data shown in Figure 1.10.

Figure 2.23 shows a comparison between the computation of the estimated wind speed and that measured by an anemometer installed on the buoy. Included also in the figure is the significant wave height computed by Equation (2.3-7) using the estimated wind speed and distance. As seen, the overall agreement between measured and estimated values is reasonably good in view of the fact that the estimation is based solely on atmospheric pressure gradients.

The estimated radius (distance) of the maximum wind is slightly less than the measured radius (35 *versus* 38 km), so is the estimated magnitude of the maximum wind speed less than the measured speed (31.3 *versus* 35.0 m/s). The same trend can also be observed in the magnitude of significant wave height, i.e. the estimated height is 8.38 m, while

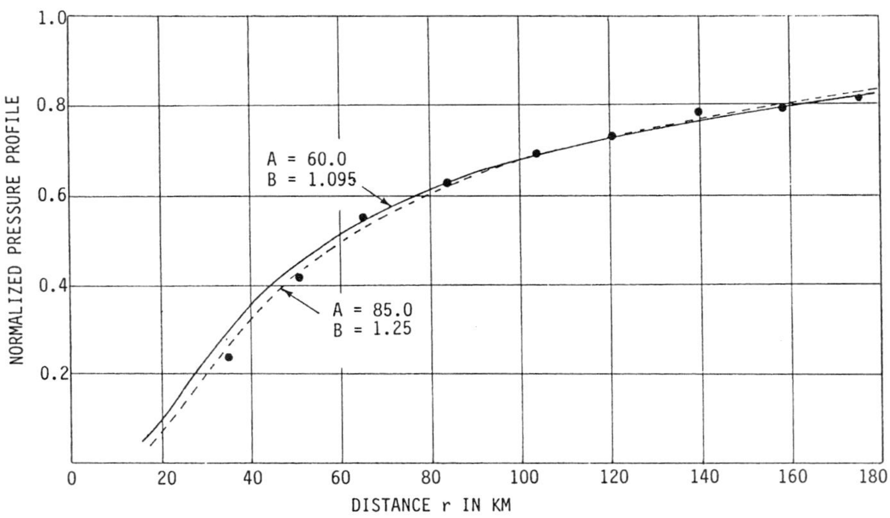

Figure 2.22 – Computed pressure profile as a function of distance between hurricane ELOISE and NOAA buoy EB10.

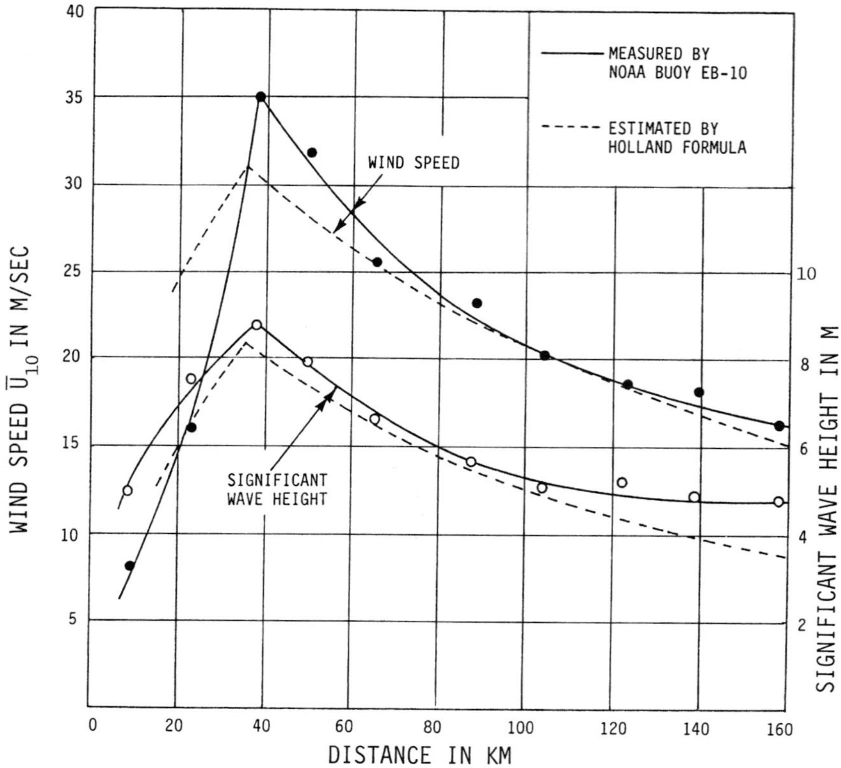

Figure 2.23 – Comparison between estimated and measured wind speed and significant wave height as a function of distance from hurricane ELOISE and NOAA buoy EB10.

the measured height is 8.78 m. This shortcoming, however, may be improved if computations are made using accurate meteorological isobars.

It is seen in Figure 2.23 that measured wind speed and significant wave height are both larger than the estimated values for distances greater than 130 km. This discrepancy, however, may be attributed to the significant change in the direction of advance from NNW to NE of hurricane ELOISE which took place when the hurricane center was approximately 120–200 km from Buoy EB-10. During this course change, the magnitude of wind speed and significant wave height measured by Buoy EB-10 are both by and large constant for 6 hours at 16–18 m/s, and 5.0 m, respectively. After the course change was stabilized, wind speed and significant wave height both started to increase as indicated in Figure 2.23. Thus, the trend of decreasing wind speed and significant wave height with increasing distance as indicated by the results of the computations appears to be legitimate.

Modal frequencies are estimated at distances of 104, 65, and 35 km from the hurricane center by Equation (2.3-3). At a distance of 35 km from the hurricane center the estimated wind speed becomes maximum as shown in Figure 2.23. Wave spectra $S(f)$ can be estimated at these locations by applying Equation (2.2-23).

Figure 2.24 shows a comparison between the estimated and measured wave spectra at locations of 104 and 65 km from hurricane ELOISE in her direction of advance, while Figures 2.25 shows comparison at the radius of maximum wind speed of 35 km. As seen in these figures, the overall agreement between estimated and measured spectra is satisfactory. Due to inaccuracy in estimating pressure gradients near the hurricane center from data obtained by the buoy, some discrepancy between measured and estimated spectra is unavoidable. The results of computations, nevertheless, demonstrate that hurricane-generated wave spectra at a specified distance from the hurricane center can be estimated from the knowledge of pressure gradients. It is highly recommended that the evaluation be carried out by employing accurate meteorological isobars and that the parameters involved in Holland's method be determined independently for distances close to or far from the hurricane center.

In summary, the procedure for estimating the wave spectrum at a specific site located in the approaching path of a hurricane in deep water from knowledge of the atmospheric pressure is as follows.

The basis for the estimation is the gradient wind velocity evaluated using Holland's formulation for normalized pressure given in Equation (2.3-11). For this,

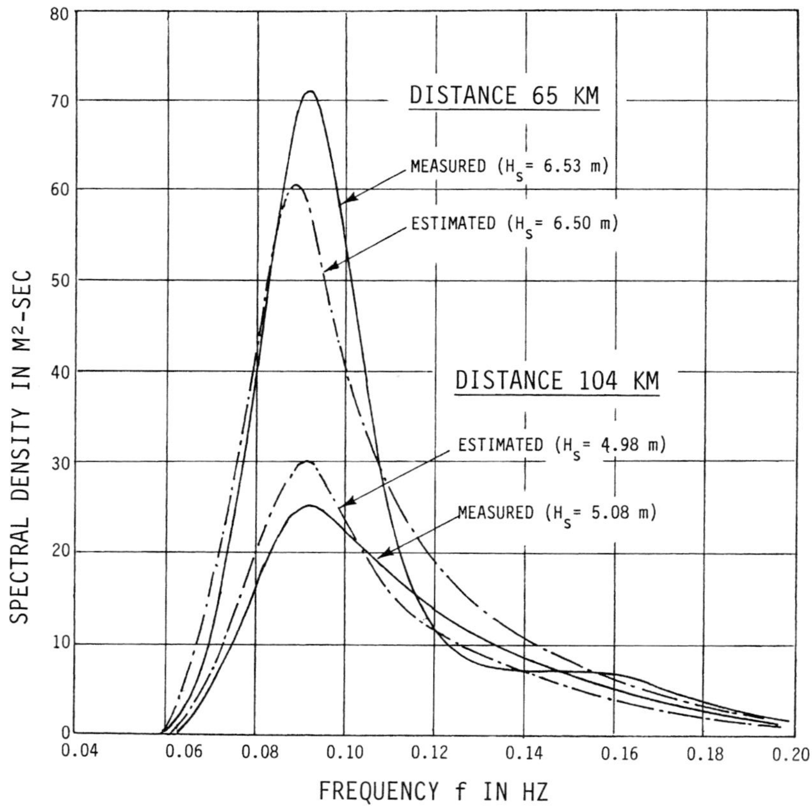

Figure 2.24 – Comparison between estimated and measured wave spectra at locations 104 and 65 km from hurricane ELOISE in her moving direction.

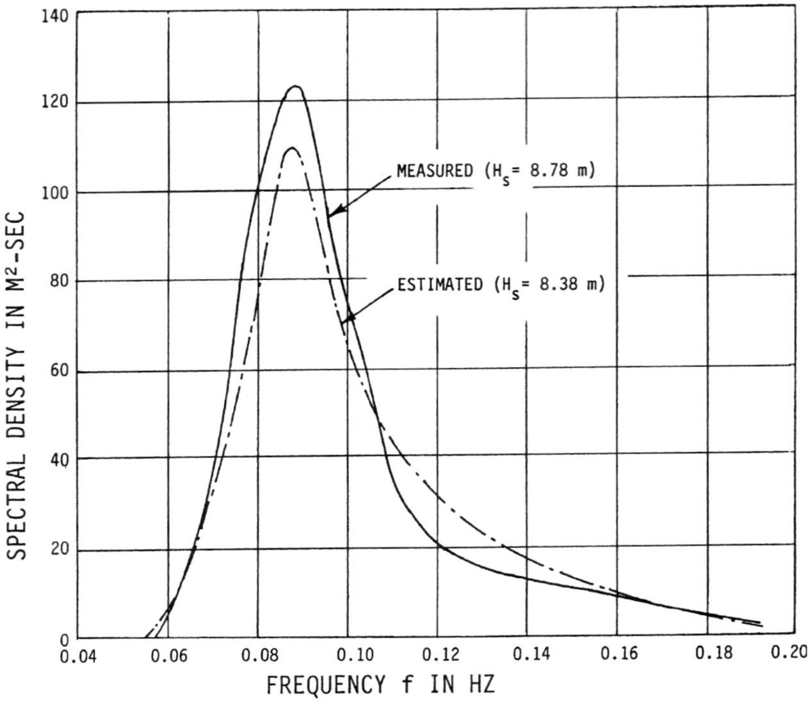

Figure 2.25 – Comparison between estimated and measured wave spectra, hurricane ELOISE at radius of maximum winds speed.

simultaneous information of (a) hurricane central pressure, (b) ambient (neutral) pressure, (c) pressure at the specific site, and (d) distance to the hurricane center are required. It is noted that in order to accurately present the normalized pressure at the site by Equation (2.3-11), the above information should be obtained several times during the hurricane's approach until its center is within a distance of no more than 40–50 km from the site. This implies that, in practice, estimation of wave spectra (including spectra at an earlier stage) is feasible subsequent to the hurricane center approaching to within about 40–50 km from the site.

Upon evaluating the parameters representing the normalized pressure profile, the gradient wind velocity for a specified distance to the hurricane center is estimated by Equation (2.3-16), and the mean wind speed is obtained therefrom by dividing by the gust factor. A gust factor of 1.23 obtained from analysis of hurricane KATE appears to be reasonable, since this is equivalent to multiplying by gradient wind velocity by 0.813 which is very close to the factor of 0.80 as suggested by Powell. From knowledge of the mean wind speed, U, for a specified distance, X, to the hurricane center the significant wave height and modal frequency of the wave spectrum can be evaluated by Equations (2.3-7) and (2.3-3), respectively. The wave spectrum can then be obtained from Equation (2.2-23). The spectrum in terms of frequency ω can be evaluated by Equation (2.2-24) by converting the frequency from f to ω.

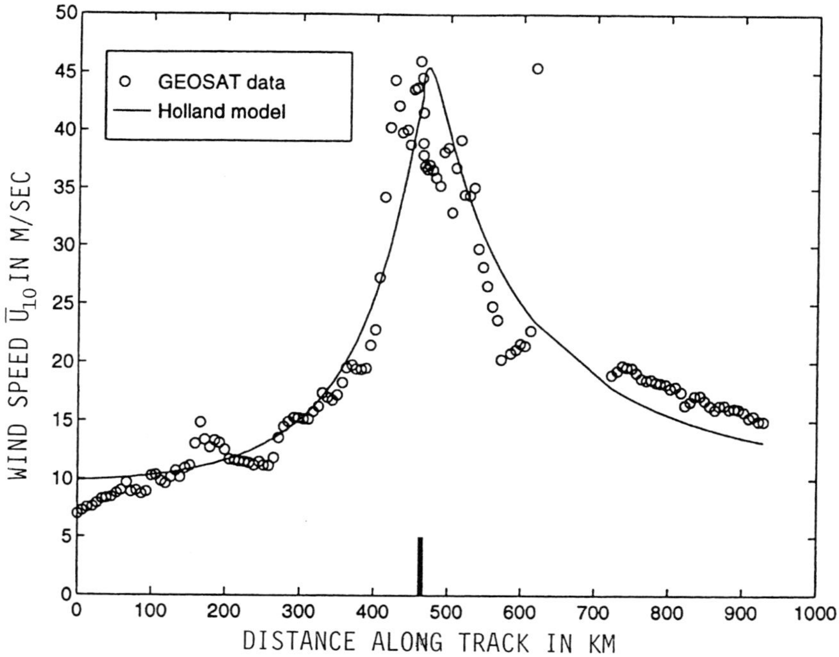

Figure 2.26 – Comparison between wind speed $\bar{U}_{10}$ as a function of distance along the track of a hurricane as observed by GEOSAT satellite and predicted relationship computed by Holland's model (Young and Burchell, 1996).

In reference to Holland's formulation, it is of interest to note that Young and Burchell (1996) evaluate the wind speed at 10 m height as a function of distance along the track of a hurricane by using Holland's model and compare it with the relationship obtained from analysis of data observed by the GEOSAT satellite. The mean wind speed at 10 m height is estimated by multiplying the gradient wind velocity by a factor of 0.80 as proposed by Powell. The agreement between them is excellent as shown in Figure 2.26.

2.4 Directional characteristics of hurricane-generated seas

The wave spectra of hurricane-generated seas discussed in Sections 2.1–2.3 represent the accumulation of energy of all waves measured at a certain location in the ocean and the direction in which the energy is traveling has not been discussed. In hurricane-generated seas, the traveling direction of the prime wave energy depends on the location of the hurricane center relative to the site.

Prior to discussing the wave directional characteristics of hurricane-generated seas, let us first consider ordinary wind-generated random waves in the ocean in which the winds are continuously blowing from a certain direction. Wave energy, in this case, does not necessarily propagate in the same direction as the wind; instead, the energy usually spreads within some frontal range of angles; although the major part of the energy propagates in the wind direction.

Usually, this spreading status of wave energy does not drastically change with time during the storm.

Wave directional characteristics in hurricane-generated seas, however, must be considered from an entirely different view point, since the hurricane-associated wind is a great counter-clockwise (in the Northern hemisphere) revolving vortex system with diameter on the order of 200 km and the system is advancing in a particular direction. Hence, when the hurricane center is approaching a specific site in the ocean, the initial wind direction at the site depends on whether the hurricane is coming on the right, on the left or directly straight ahead to the site; additionally, this initial wind direction changes its heading continuously with time. Thus, the directional characteristics of hurricane-generated seas are extremely complicated; it is first necessary to clarify the direction where the prime wave energy is moving with reference to the relative location of the hurricane center to the site, and then examine the energy spreading condition around the prime energy movement at each location.

Several studies on the directional characteristics of hurricane-generated seas have been carried out. These include Cardon and Pierson (1975), Forristall *et al* (1978), Black (1979), and Forristall and Ward (1980), among others. Although information from observed data by employing wave gauges along with current meters is extremely useful, there are some significant limitations in these studies to adequately model the actual phenomena of interest because of reasons mentioned in the previous paragraph.

Valuable information was obtained by the NOAA airborne synthetic aperture radar (SAR) system aboard aircraft which flew over hurricanes EMMY, FRANCES, and GLORIA in 1976. In these flights, the directional distribution of wave fields was obtained through observations made over most of the area within a radius of 150 km around the center of the eye. The SAR system was also used as one of the five remote sensing instruments on board the SEASAT launched in 1978, and some information on wave characteristics associated with hurricane IVA was acquired during its revolution around the earth (Gonzalez *et al*, 1982).

Some results of analysis of data obtained during hurricane JOSEPHINE are presented by Holt and Gonzallez (1986), and results of data obtained during hurricane GLORIA are reported by Elachi, Thompson, and Weissman (1977) and King and Shemdin (1978). The results obtained by Elachi *et al* show that the maximum surface wind was 60 m/s (117 knots), and the advancing speed of hurricane GLORIA was 28 km/h (7.78 m/s, 15.1 knots) when the observation was made. The wind field was stronger on the right side of the hurricane as expected in the Northern hemisphere which resulted in substantial asymmetry in the wave field.

Figure 2.27 taken from Elachi's publication shows an interesting wave pattern under a hurricane. Due to the inherent resolution of the system, only waves of length greater than 75 m were imaged by the sensor. In the figure, the wave fronts are shown as pairs of parallel lines. The separation between the pairs of lines is proportional to the wavelength, and the numbers indicate the wavelength in meters. Several interesting patterns can be seen in the figure. These are

a) Around the area of maximum wind (about 60 km from the eye), the wave direction is close to the wind direction.

b) Waves in the eye region are short-crested which is typical of a local wind-generated sea.

c) Waves in the region forward of the moving hurricane eye appear to be propagating in the radial direction, which is roughly perpendicular to the local wind. Wavelengths in this

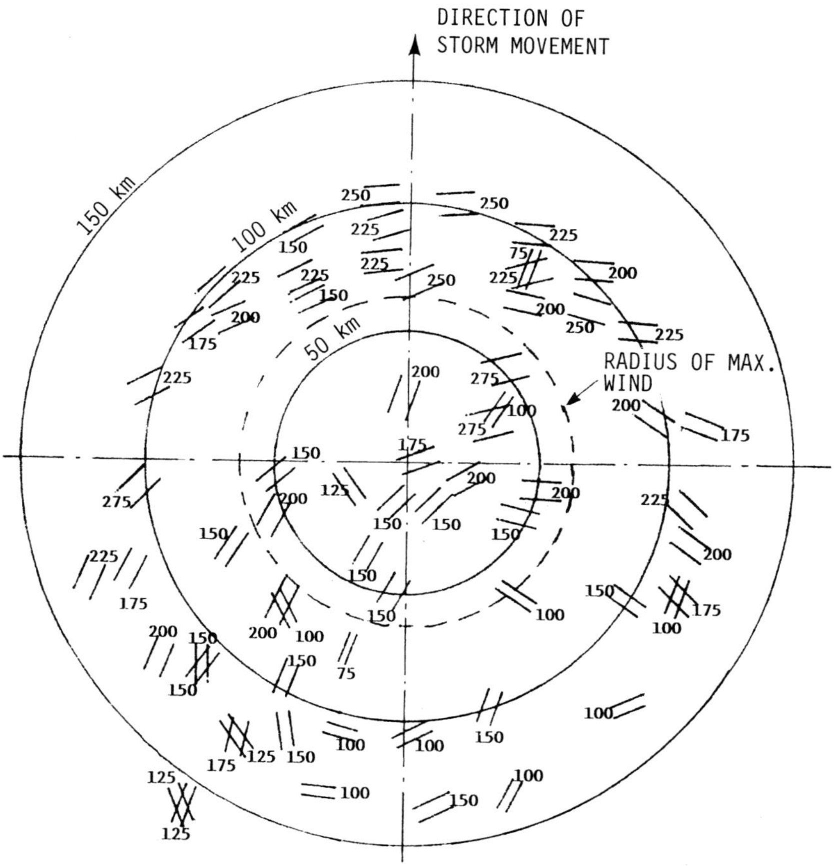

Figure 2.27 – Wave patterns during hurricane GLORIA (Elachi, Thompson, and Weissman, 1977).

region are all large and traveling ahead of the hurricane as swell. (This subject is discussed in detail in Section 2.1.)

d) Waves in the rear half region are of shorter length (and thereby of smaller periods, especially in the left rear quadrant).

Analysis carried out by King and Shemdin (1978) derive results similar to those obtained by Elachi *et al*, and they stress the asymmetry in the wave field resulting from the forward motion of the hurricane; i.e. in the northern hemisphere, the counter-clockwise winds in a hurricane blow faster on the right hand side of the moving hurricane than on the left due to the added speed of the storm. This causes larger waves to be generated in this area and these waves tend to travel with the storm. Comparison between the dominant wave number vectors and wind directions obtained in hurricane GLORIA given by King and Shemdin is shown in Figure 2.28. It is of interest to see that the wind and wave directions are opposite on the left

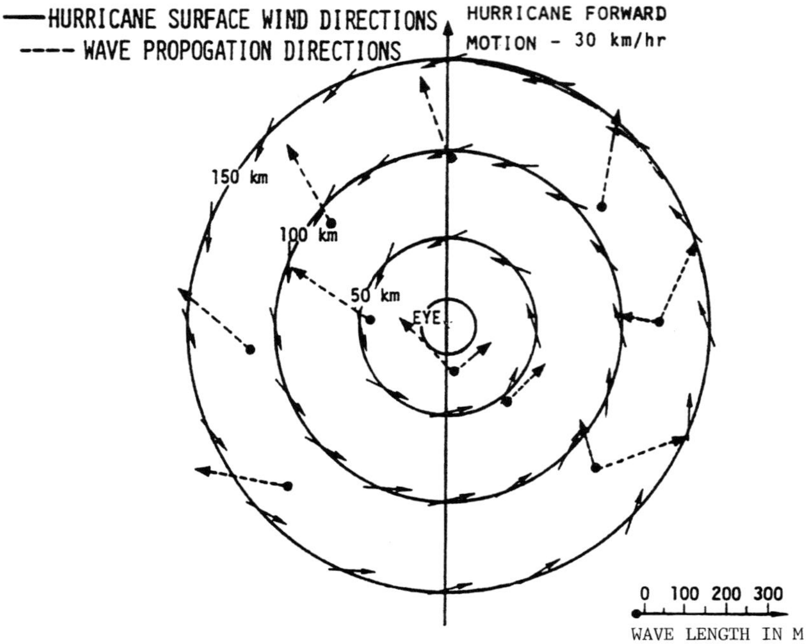

Figure 2.28 – Dominant wave number vector compared to wind direction in hurricane GLORIA (King and Shemdin, 1978).

side of the moving hurricane, the same trend as observed in Figure 2.27. Although the results of the studies by Elachi *et al* and King and Shemdin are limited in clarifying the direction of prime wave energy with respect to the relative location of the hurricane center to a specific site in the ocean, they demonstrate that the observations with the help of aircraft flights over hurricanes appear to be most beneficial in that the aircraft has freedom to control the locations where the radar imagery can determine the wave direction and wavelength within a specified radial distance around the eye.

Recently, an extensive field study on directional wave characteristics in hurricane-generated seas was carried out by a group of researchers by employing the NASA scanning radar altimeter (SRA) system carried on board the NOAA hurricane research aircraft during hurricane BONNIE (Wright *et al*, 2001). The aircraft spent more than 5 hours within 180 km of the eye and obtained directional wave spectra in all quadrants of the hurricane's inner core over open water (see Figure 2.29).

Exhaustive analyses were made on the measured data, the results of which are presented in 45 wave directional spectra along with the prime wind direction, providing a thorough picture of the wave energy spreading properties of hurricane-generated seas.

The severest sea obtained during the measurements had a significant wave height of 10.9 m. Figure 2.30 shows three wave directional spectra taken from Wright *et al*'s publication, all of which are of the severest sea conditions; two of them are close to the hurricane center (67 and 83 km from the center) with significant wave height

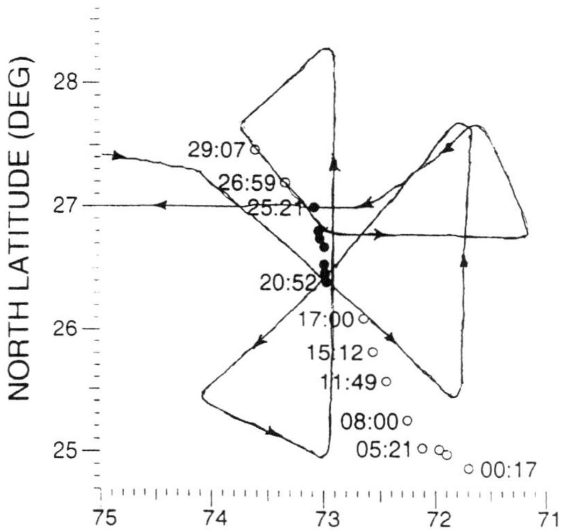

Figure 2.29 – Eye position of hurricane BONNIE and flight course taken by NOAA Research aircraft/US Air Force Reserve hurricane aircraft (Wright et al, 2001).

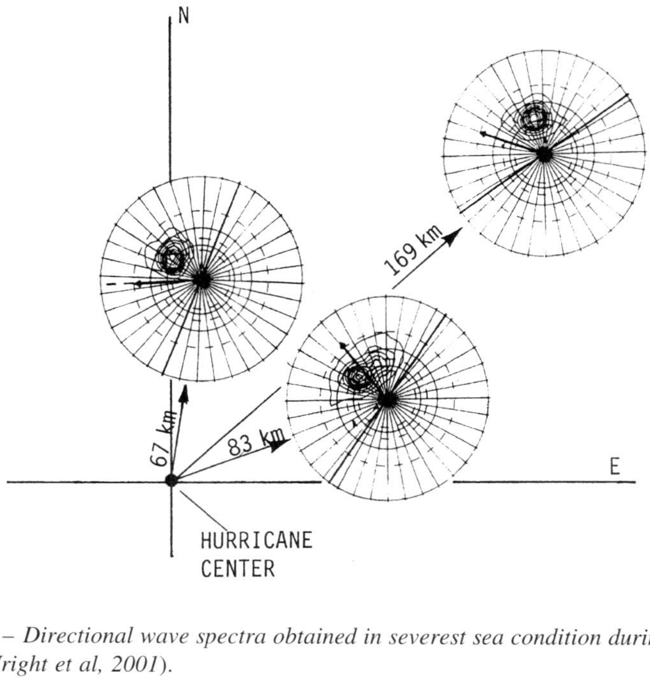

Figure 2.30 – Directional wave spectra obtained in severest sea condition during hurricane BONNIE (Wright et al, 2001).

of 10.8 and 10.9 m, respectively; the third is quite a distance from the center (169 km) with significant wave height of 10.8 m. The advancing direction of the hurricane is slightly off north toward east. Note that the three records were not taken simultaneously. The heavy arrow in each directional distribution indicates the prime wind direction. As expected, severe sea states are observed in the right forward quadrant of the moving hurricane and the energy spreading of these severe seas is rather narrow, approximately 20° on both sides of the direction in which the major wave energy propagates.

Figure 2.31 shows the primary wave field of hurricane BONNIE, in which the radials extend in the wave propagation direction. The radial length is proportional to the wavelength and the width is proportional to significant wave height so that the aspect ratio is an indication of wave steepness; a wide rectangle indicates steep waves. The largest waves (300 m wavelength with 10 m high swell) are observed in the right forward quadrant.

Wright *et al* further evaluate the effective wind for generating waves propagating toward the site. For this, at each location, for every 10° azimuth, the wind effect over a 300 km fetch is summed in 10 km increments. At the 31st position along each radial direction an "effective wind" for generating waves propagating toward the site is calculated by applying the Hurricane Research Division (HRD) surface wind analysis. Figure 2.32 shows the evaluated effective wind directional patterns overlaid on the wave directional spectrum at six various locations. Included also in the figure is the identification of these six locations relative to the hurricane center. As seen, the direction of the maximum of the wind pattern agrees well with the observed propagation direction of the dominant waves.

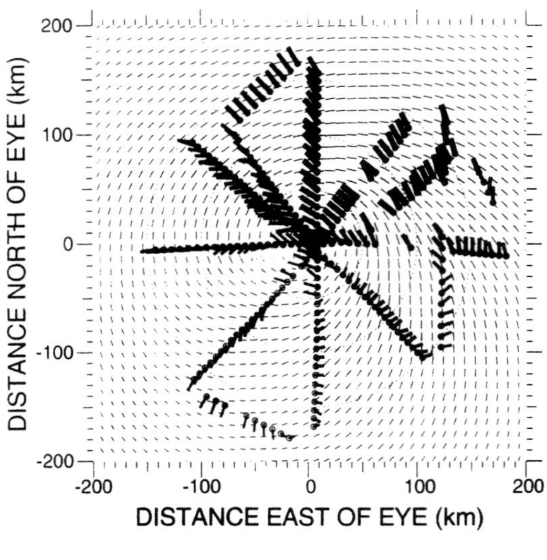

Figure 2.31 – Hurricane BONNIE primary wave field (Wright et al, 2001).

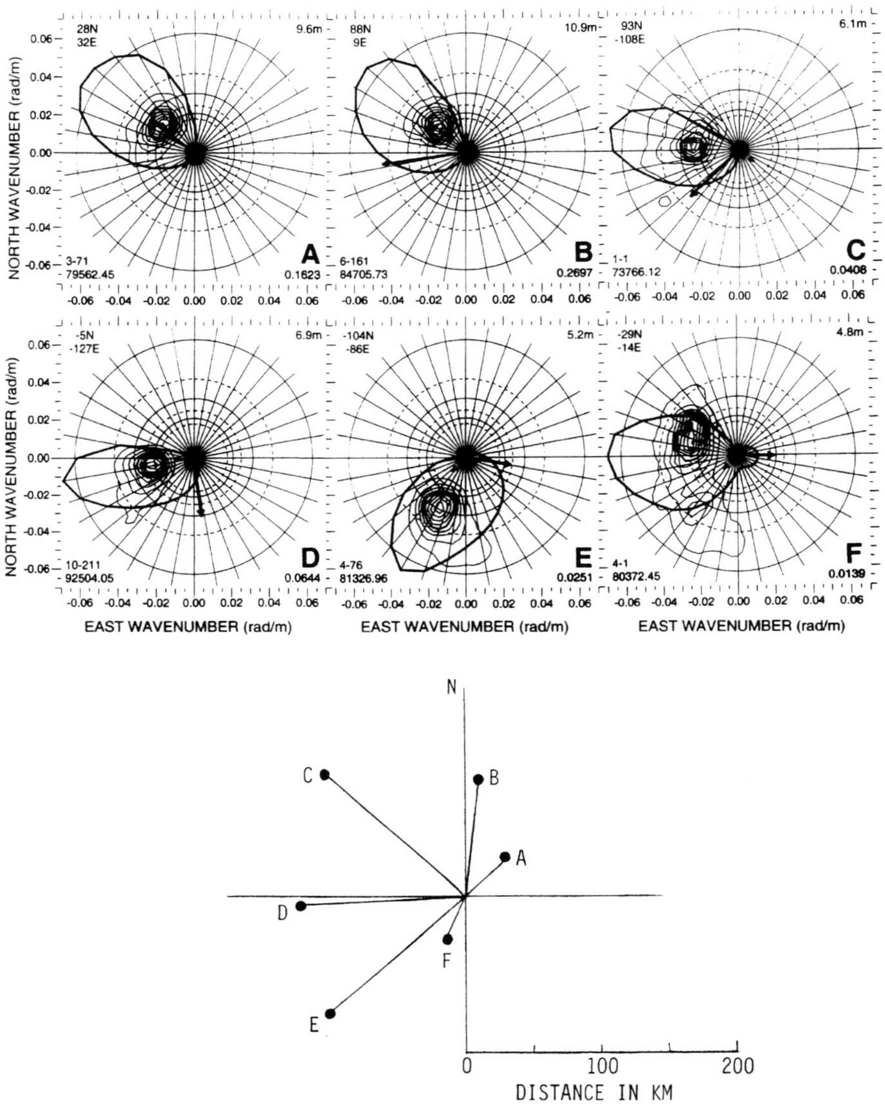

Figure 2.32 – Wind-weighting pattern overlaid on six spectra for various positions of hurricane BONNIE (Wright et al, 2001).

Furthermore, Walsh *et al* (2002) carry out an extensive study on wave directional characteristics from data obtained when hurricane BONNIE was moving over the sea in finite water depth before landing, and they find that the wave directional characteristics are nearly the same as those obtained earlier in deep water.

3

Transformation of wave spectra with the advance of a hurricane from deep to finite water depth

In this chapter the reduction of sea severity (transformation of wave spectrum) when a hurricane is advancing from deep to finite water depth is discussed. Information on wave spectra in finite water depth is extremely important in that the majority of structures in petroleum engineering are located on the continental shelf. As mentioned in Chapter 2, the severity of a hurricane-generated sea at a specified location in the ocean depends on the sea condition prior to the arrival of the hurricane at the site and on the magnitude of wind speed and distance of the site from the hurricane center. We may estimate the wave spectrum by applying the methodology discussed in the preceding chapter as long as the hurricane is moving over deep water.

The wave spectrum, however, undergoes a reduction of energy when a hurricane reaches the continental shelf where the water depth affects the wave characteristics. It is commonly known that the water depth at which the reduction of energy is initiated can be roughly estimated as one-half of the wavelength corresponding to the frequency where the spectrum peaks. More precisely, the limiting depth where reduction of energy initiates can be evaluated by applying Kitaigorodskii's transformation factor as will be discussed in detail in this chapter.

As the hurricane further advances toward the shoreline, the wave energy reduces due to shoaling and the wave profile begins to deviate from that observed in deep water. In other words, waves are transformed from a Gaussian to a non-Gaussian random process. Additionally, a substantial reduction of energy will occur due to breaking. In evaluating the energy loss in the wave spectrum associated with wave breaking, consideration of non-Gaussian wave characteristics is required. With continued advancement towards the shoreline, a significant reduction of wave energy occurs due to the wave-bottom interaction process in shallow water.

Since finite water depth covers a wide range of the ocean from the border of the continental shelf to the shoreline, this region is divided into two areas in the following discussion on wave spectra; namely, the continental shelf and nearshore area. The continental shelf covers the area from the site where reduction of energy in the wave spectrum initiates to a location where the wave breaks (breaking point). The reduction of wave energy is identified by evaluating the value of Kitaigorodskii's transformation factor at which it deviates from unity. The nearshore area is defined as that from the breaking point to the calm water shoreline. Wave spectra in these two regions are discussed in Sections 3.1 and 3.3, respectively, while Section 3.2 is devoted to a discussion of the energy loss associated with breaking at a transition point between the two regions.

A great number of studies on wave spectra in finite water depth have been made analytically and experimentally as well as through field observations. Just a short list of studies include Kitaigorodskii, Krasitskii, and Zaslavskii (1975), Thornton (1977), Knowles (1982), Huang *et al* (1983), Young (1988a,b), Miller and Vincent (1990), Holhuijsen, Booij, and Ris (1993), etc. Among others, Kitaigorodskii *et al*'s contribution provides significant insight to many studies on wave spectra in finite water depths.

3.1 Wave spectra in finite water depth on the continental shelf

A method for evaluating the transformation of wave spectra from deep to finite water depth was analytically developed by Kitaigorodskii, Krasitskii, and Zaslavskii (1975), Kitaigorodskii (1983) and the validity of their methodology was proved through many examples of field data. In particular, their transformation method was applied to the JONSWAP wave spectrum; the result, called the TMA spectral formulation, has been widely used for the design of marine systems in finite water depth (Bouws *et al*, 1985). This transformation method can also be applied for estimating sea severity associated with hurricane-generated seas on the continental shelf from knowledge of the wave spectrum in deep water. Kitaigorodskii *et al*'s transformation formula is as follows

Assuming that the effect of viscosity of water and surface tension on wave motion can be neglected, Phillips (1958, 1966) derive the equilibrium range of a deep water wave spectrum in the following form

$$S(\omega) = (\text{constant})g^2\omega^{-5} \tag{3.1-1}$$

Kitaigorodskii *et al* extend Phillips' concept to waves in finite water depth by presenting the equilibrium range of a wave spectrum in finite water depth in the form of

$$S(\omega) = (\text{constant})g^2\omega^{-5}\Phi(\omega_h) \tag{3.1-2}$$

where $\Phi(\omega_h)$ is the non-dimensional function with a parameter ω_h.

$$\omega_h = \omega\sqrt{h/g}$$

where h is the water depth.

Further, Kitaigorodskii *et al* analyze waves in finite water depth obtained by Dreyer and find that the equilibrium range of shallow water waves can be presented by a wave number spectrum given in the form of

$$S(k) = (\text{constant})k^{-3} \tag{3.1-3}$$

where the wave number k is associated with the following dispersion relationship for gravity waves given by

$$\{\omega(k)\}^2 = gk\tanh(kh)$$

Let us write the wave number as a function of frequency ω and water depth as,

$$k(\omega) = \frac{\omega^2}{g} U(\omega_h) \tag{3.1-4}$$

where $\omega_h = \omega\sqrt{h/g}$, and thereby we have

$$kh = \omega_h^2 U(\omega_h) \tag{3.1-5}$$

Then, from the dispersion relationship and Equation (3.1-5), we can determine the function $U(\omega_h)$ from the following equation

$$U(\omega_h)\tanh(\omega_h^2 U(\omega_h)) = 1 \tag{3.1-6}$$

Next, in order to determine the function $\Phi(\omega_h)$ in Equation (3.1-2), the wave number spectrum in finite water depth given in Equation (3.1-3) is expressed in terms of frequency as

$$S(\omega, h) = [S(k)]_{k=(\omega^2/g)U(\omega_h)} \left| \frac{\partial k}{\partial \omega} \right| \tag{3.1-7}$$

Hereafter, the wave spectrum in finite water depth may be written as $S(\omega, h)$ where h is the water depth. With the aid of Equations (3.1-4) and (3.1-5), Equation (3.1-7) can be written as

$$S(\omega, h) = (\text{constant}) g^2 \omega^{-5} \{U(\omega)\}^{-2} \left[1 + \frac{2\omega_h^2 U(\omega_h)}{\sinh(2\omega_h^2 U(\omega_h))} \right]^{-1} \tag{3.1-8}$$

Thus, from comparison of Equations (3.1-2) and (3.1-8), we have

$$\Phi(\omega_h) = \{U(\omega_h)\}^{-2} \left[1 + \frac{2\omega_h^2 U(\omega_h)}{\sinh(2\omega_h^2 U(\omega_h))} \right]^{-1} \tag{3.1-9}$$

The function $\Phi(\omega_h)$ so derived is called Kitaigorodskii's transformation factor, and it plays an extremely important role for estimating the wave spectrum in finite water depth from knowledge of the wave spectrum in deep water.

In Equation (3.1-9), $\Phi(\omega_h)$ becomes unity for $\omega_h \to \infty$. On the other hand, $\Phi(\omega_h)$ becomes approximately $\omega_h^2/2$ for $\omega_h \to 0$ which leads to the spectrum in the form of

$$S(\omega) = (\text{constant}) gh\omega^{-3} \tag{3.1-10}$$

Kitaigorodskii *et al* analyze waves in finite water depth obtained by Dreyer and found that the equilibrium range of shallow water waves can be presented by Equation (3.1-10).

In practice, the numerical computation of Equation (3.1-9) shows that the transformation factor $\Phi(\omega_h)$ is small for lower frequencies, while it is close to unity for high frequencies as shown in Figure 3.1. This implies that, for a given water depth, the reduction in wave energy is much larger for the lower frequencies of the deep water wave spectrum than for the higher frequencies. It is noted that the wave spectrum in a water depth, h, is often written as the product of the wave spectrum in deep water, $S(\omega, \infty)$, and the transformation factor as follows

$$S(\omega, h) = S(\omega, \infty) \left[\frac{(k(\omega, h))^{-3} \dfrac{\partial k(\omega, h)}{\partial \omega}}{(k(\omega, \infty))^{-3} \dfrac{\partial k(\omega, \infty)}{\partial \omega}} \right]_{\omega_h = \omega\sqrt{h/g}} \tag{3.1-11}$$

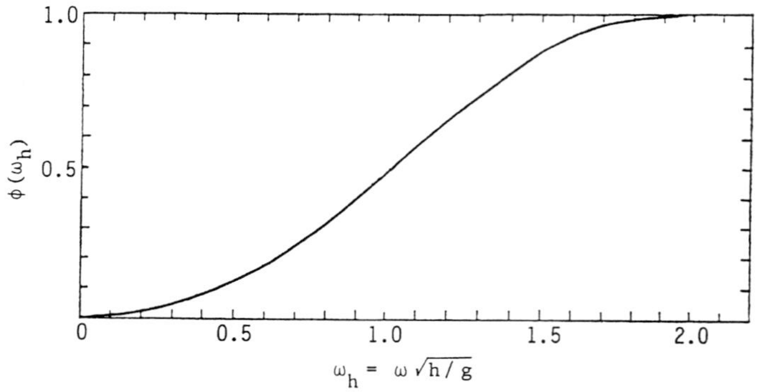

Figure 3.1 – Transformation factor as a function of dimensionless frequency ω_h (Kitaigorodskii, Krasitskii, and Zaslavskii, 1975).

The validity of the transformation factor for estimating the wave spectrum in finite water depth was proved by Bouws *et al* (1985) through a study using over 2500 wave spectra measured at various water depths in three wave measurement projects, namely, MARSEN, ARSLOE, and TEXEL, and found that the transformation factor is not restricted to the equilibrium range but appears to be valid across the entire frequency range of the spectrum. The transformation factor is then applied to the JONSWAP wave spectral formulation; the result is the well known TMA spectral formulation which has been widely used for the design of marine systems in finite water depth.

A limitation in applying the transformation factor for wave spectra in finite water depth is that the application is valid only to any water depth for which no wave breaking is involved. It is also noted that the transformation factor is a function of frequency. Therefore, wave spectra in deep water having the same severity (significant wave height), but different shapes result in different sea severities (significant wave heights) in a finite water depth.

Let us evaluate hurricane-generated wave spectra at various water depths by applying the transformation factor to a spectrum obtained in deep water. As an example, Figure 3.2 shows a wave spectrum obtained from data measured in deep water by NOAA Buoy 42003 (25.9N, 85.9W) in the severest sea condition encountered during hurricane KATE. The track of hurricane KATE is shown in Figure 3.3. Hurricane KATE originated north of the Virgin Islands in November 1984 and passed over Puerto Rico, Cuba, and entered the Gulf of Mexico. It passed nearby Buoy 42003 as shown in the figure, and finally made landfall on Crooked Island, east of Panama City Case (1986). Point A in the figure indicates the location where the effect of water depth was first recognized in the wave spectra computed at various water depths starting from deep water. This is explained below.

Assuming that the shape of wave spectra in the severest sea state shown in Figure 3.2 remains constant as long as the hurricane is moving with a constant wind speed in deep water, wave spectra on the continental shelf can be estimated by applying Equation (3.1-11). The results of the computations made for various depths from deep water (300 m) show that the transformation factor given in Equation (3.1-11) remains unity until the water depth is reduced to 120 m in this example. This implies that the effect of finite water depth initiates at this water depth (Point A in Figure 3.3) for hurricane KATE.

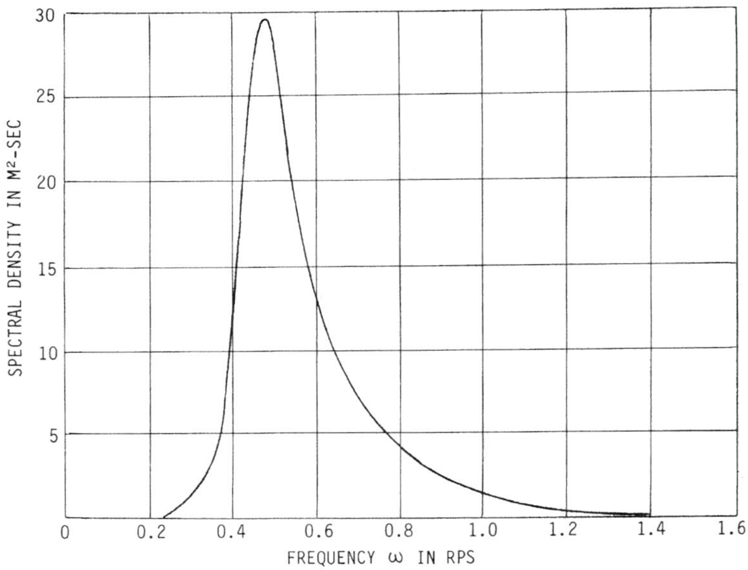

Figure 3.2 – Wave spectrum of hurricane KATE obtained from data measured by NOAA buoy 42003.

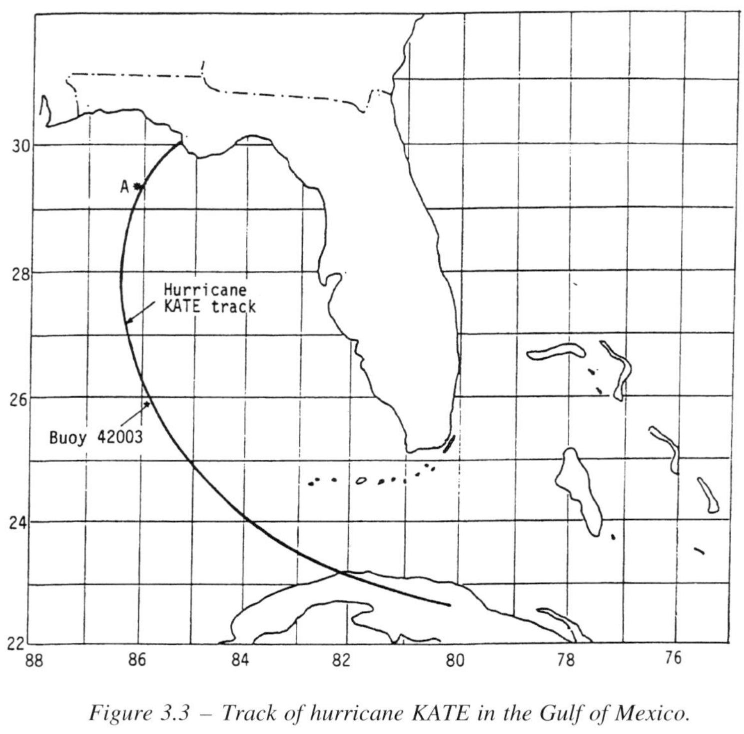

Figure 3.3 – Track of hurricane KATE in the Gulf of Mexico.

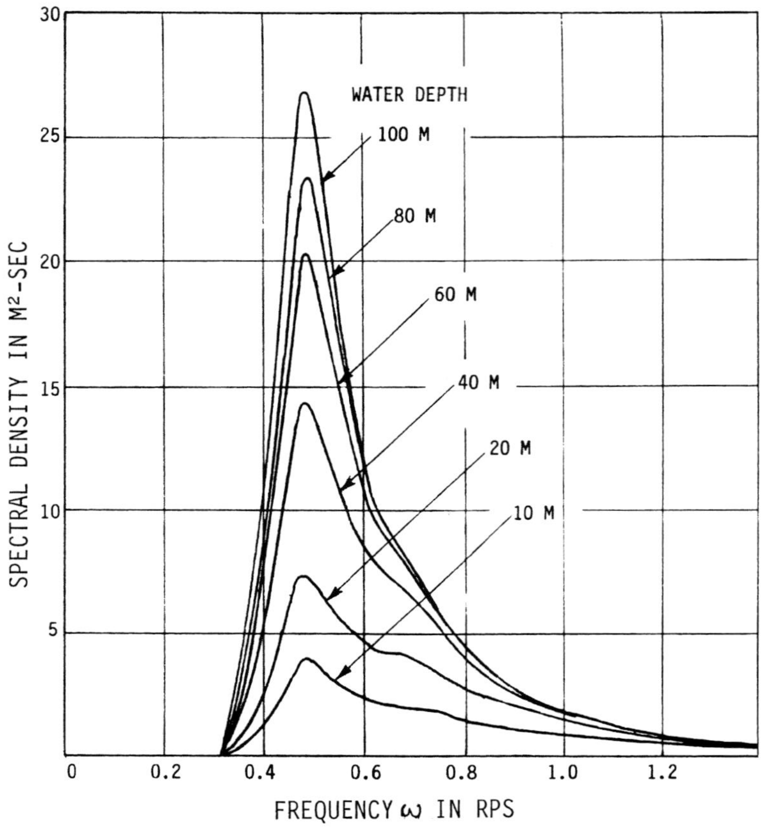

Figure 3.4 – Wave spectra of hurricane KATE for various water depths computed by applying Kitaigorodskii's transformation factor to deep water measured spectrum.

The shapes of hurricane KATE wave spectra computed for various water depths ranging from 100 to 10 m are shown in Figure 3.4. As seen, the sea severity reduces drastically with decrease in water depth.

Figure 3.5 shows the reduction of energy represented by the ratio of the area under the spectrum (variance) at a given water depth to that in deep water plotted as a function of water depth. As seen, the reduction of energy initiates at a water depth of 120 m. A 20% reduction in sea severity can be seen at 60 m water depth, a 50% reduction at 30 m, and 90% reduction at 10 m water depth in this example.

Figure 3.5 may give the impression that waves can no longer be considered linear for water depths less than 120 m in hurricane KATE. Further, since wave non-linearity results in alteration of the wave profile in general, the question arises as to the water depth at which the wave characteristics deviate from a Gaussian random process. This interesting and important subject in evaluating the non-Gaussian wave property from the wave spectrum will be discussed in detail later in Section 4.2. It may suffice here to state that deviation from deep water Gaussian wave characteristics is not appreciable until a water depth on the order of 33 m is reached for hurricane KATE.

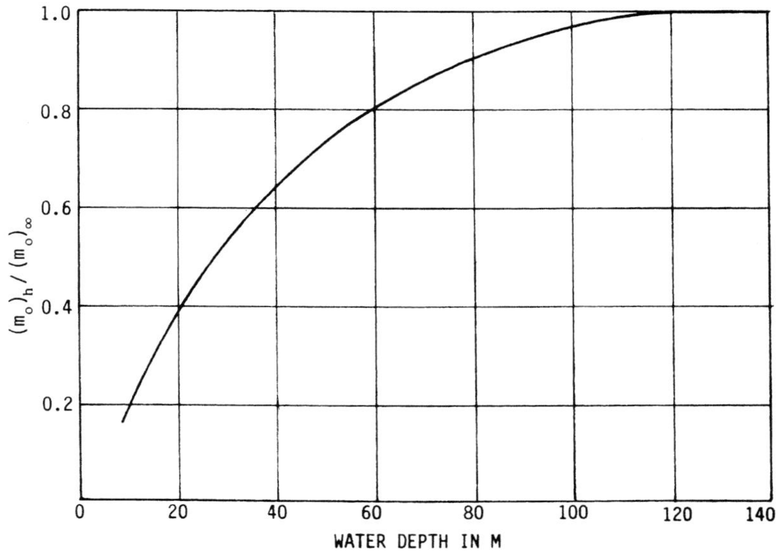

Figure 3.5 – Ratio of area under computed spectrum at a given water depth to that in deep water as a function of water depth.

3.2 Wave breaking and energy loss

3.2.1 Wave breaking

In this chapter, wave breaking refers to that observed in finite water depth. In general, waves traveling landward on the continental shelf break in the vicinity of a certain water depth whenever a momentarily high wave crest reaches an unstable condition; this can be characterized as a function of wave height and length (period) for the given water depth. The location where the breaking phenomenon takes place is called the breaking point. It is not a fixed location in random seas, but it is possible to estimate the breaking point by evaluating the frequency of occurrence of breaking as well as the energy loss associated with breaking. A detailed discussion on this subject will be given in Section 3.2.3.

It is noted that breaking waves are often classified as spilling, plunging, and surging breakers. It is difficult, however, to apply such classification to random waves; some waves break by plunging while others by spilling, depending on sea severity and individual wave height and length (thereby period) relationship along with the water depth. Furthermore, it should be noted that the characteristics of waves near the breaking point are categorized as non-Gaussian random waves. This must be considered while predicting the occurrence of breaking and the loss of wave energy as well.

Although many studies have been carried out on wave breaking in finite water depth (in particular, in the shoaling zone) from different approaches and view points, the majority of these studies deal with breaking of regular waves, and only a limited number of studies have been made on breaking of random waves. Kimura and Iwata (1978) conclude from the results of their experimental study on two different bottom slopes that breaking characteristics of random waves are consistently different from those of periodic waves.

As to the wave condition under which breaking takes place in finite water depth, Komar and Gaughan (1972) derive the following criterion on wave height and find that the criterion satisfactorily agrees with the results of analysis of experimental work as well as field data

$$H_b = 0.39 g^{0.20} (H_0^2 T)^{0.40} \tag{3.2-1}$$

where H_b is the breaking wave height, H_0 the wave height in deep water, and T the wave period.

Weishar and Byrne (1978) carry out field observations of breaking wave behavior at Virginia Beach and find that the incidental wave height at which a wave breaks also agrees well with the criterion given by Komar and Gaughan. Furthermore, Iwata and Sawaragi (1982) demonstrate that the incidental breaking wave height also agrees well with the Komar and Gaughan's criterion, irrespective of whether breaking is of the spilling or plunging type, in their experimental study on breaking of random waves over three different bottom slopes.

Another criterion often considered for breaking waves is the ratio of breaking wave height, H_0, and the water depth at breaking, h. This ratio is denoted by γ_b. By applying the Airy wave theory, Komar and Gaughan show that $\gamma_b = 1.42$ from Equation (3.2-1). Komar's experimentally measured value of γ_b, however, ranges from 0.7 to 1.1, while Munk (1949) derive $\gamma_b = 0.78$ in his analytical study based on solitary wave theory.

Kaminsky and Kraus (1993) compile 17 independent sets of experimental studies on breaking waves obtained by many researchers and find that γ_b can best be estimated by

$$\gamma_b = 1.20 I^{0.27} \tag{3.2-2}$$

where $I = m\sqrt{H_0/L_0}$ is defined as the Iribarren number or surf-similarity parameter, $m = \tan\theta$ representing the bottom slope in the nearshore area, and H_0, L_0 are wave height and length, respectively, in deep water.

Further, Kaminsky and Kraus derive the following formula for the water depth where breaking may occur.

$$h_b = 0.30\, m^{-0.25} (H_0/L_0)^{-0.23} \tag{3.2-3}$$

where h_b is the water depth where wave breaks.

Note that these criteria on the wave breaking point are given as a function of wave height and wavelength (or period) in deep water which are randomly changing in space and time; hence, it is difficult to determine these values in practice.

On the other hand, the following Miche's (1951) criterion is often considered for evaluating the breaking characteristics of waves in finite water depth

$$H \geq 0.142 L \tanh(2\pi h/L) \tag{3.2-4}$$

where h is the water depth, H wave height, and L wavelength.

The validity of the relationship given in Equation (3.2-4) may be seen in the results of the experimental study by van Dorn (1978). Although Equation (3.2-4) is for periodic waves of constant form, Battjes and Janssen (1978) apply the criterion for energy loss associated with breaking for irregular waves by truncating the Rayleigh density function at the maximum height given by Equation (3.2-4). Their work on a dissipation model for random breaking waves is further calibrated by Battjes and Stive (1985).

Recently, a method for evaluating the breaking characteristics of non-Gaussian waves in finite water depth was developed based on Equation (3.2-4) (Ochi and Malakar, 2001). By applying their method, it is possible to find the location where wave breaking as well as the magnitude of energy loss in the wave spectrum takes place when a hurricane is approaching the shoreline. These subjects will be discussed in Sections 3.2.2 and 3.2.3.

3.2.2 Frequency of occurrence of wave breaking

In order to evaluate the frequency of occurrence of wave breaking in random seas of finite water depth, we may assume the wavelength L in Equation (3.2-4) to be the Stokes limiting wavelength as considered by Michell (1893). Then, Equation (3.2-4) may be written as

$$H \geq 0.027 g T^2 (\tanh kh)^2 \tag{3.2-5}$$

where k is the wave number.

It is found in experimental studies that irregular waves in deep water break at lesser wave heights than given in Equation (3.2-5) due to the unstable characteristics of irregular waves resulting in the value of the constant involved in the criterion being 0.020 for irregular waves (Ochi and Tsai, 1983). Hence, it is appropriate to consider that the unstable characteristics of irregular waves may also be present in finite water depth, and we may write the breaking criterion as

$$H \geq 0.020 g T^2 (\tanh kh)^2 \tag{3.2-6}$$

Here, wave height H and period T are random variables in irregular waves at a specified water depth h; therefore, in principle, the joint probability distribution of wave height and period applicable to non-Gaussian waves is required for estimating the probability of occurrence of wave breaking.

However, if we consider the envelope process of non-Gaussian waves, the probability of wave breaking can be evaluated by modifying the joint distribution of Gaussian waves. In other words, in a separate treatment of peaks and troughs the statistical distribution of the peak (or trough) envelope process of non-Gaussian waves with skewness less than 1.2 (almost all waves in finite water depth satisfy this condition) are analytically proved to approximately obey the Rayleigh probability distribution (Ochi, 1998). It is generally known that if the peak envelope process of random waves follows the Rayleigh probability distribution, there exist narrow-band Gaussian waves with zero mean under the envelope process. In the case of an envelope process of non-Gaussian wave amplitude, we may say there exists equivalent narrow-band Gaussian waves as illustrated in the pictorial sketch shown in Figure 3.6. Therefore, instead of the joint probability distribution of non-Gaussian waves, the joint probability distribution developed for Gaussian waves (Cavanié, Arhan, and Ezraty, 1976; Longuet-Higins, 1983) can be applied with appropriate modification.

Prior to further discussion on estimation of the probability of occurrence of breaking of non-Gaussian waves, we may write the breaking criterion in reference to the peak of non-Gaussian waves as

$$A \geq 0.010 g T^2 (\tanh kh)^2 \tag{3.2-7}$$

where A stands for amplitude. The peak envelope process follows the Rayleigh probability distribution with a parameter equal to twice the variance of the equivalent Gaussian waves (see Equation (A.2-8) in the Appendix). The variance of the equivalent Gaussian waves,

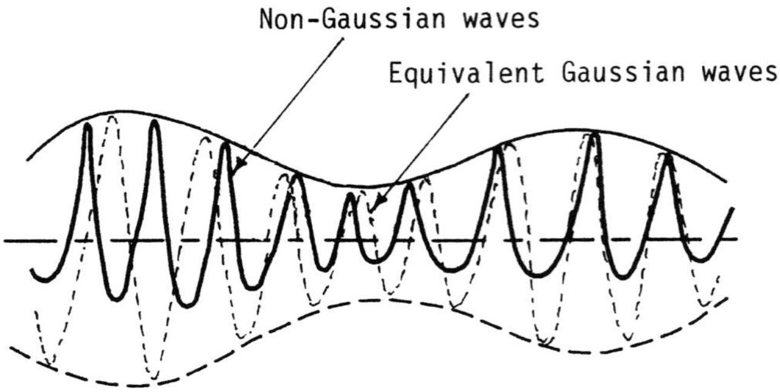

Figure 3.6 – Pictorial sketch indicating equivalent Gaussian waves under peak envelope process of non-Gaussian waves.

denoted by m_e, can be analytically derived as

$$m_e = s^2/\{1 - (2c^2/5s^2)\} \tag{3.2-8}$$

where

$$s^2 = \sigma_1^2/(1 + \lambda\mu_*)$$

$$c = \sqrt{2}(\lambda\sigma_1^2 - \mu_*)/(1 + \lambda\mu_*)$$

$$\lambda = 1.28a$$

$$\sigma_1^2 = \sigma_*^2\{1 + 2\sqrt{2/\pi}(\lambda\sigma_*) + 3(\lambda\sigma_*)^2\}$$

The parameters a, σ_*, and μ_* represent the non-Gaussian property of waves in finite water depth. The values of these parameters can be evaluated from the wave record; if the wave record is not available, they can be approximately estimated from knowledge of the variance (area under wave spectrum). Discussion on non-Gaussian waves in finite water depth is outlined in the Appendix.

Next, let us consider the joint probability distribution of wave amplitude and period. Cavanié *et al* give the dimensionless joint probability density function of wave amplitude and time interval between successive positive maxima of Gaussian waves as

$$f(\nu, \lambda) = \sqrt{\frac{2}{\pi}}\frac{\alpha^3}{\varepsilon(1-\varepsilon^2)}\frac{\nu^2}{\lambda^5}\exp\left[-\frac{\nu^2}{2\varepsilon^2\lambda^4}\{(\lambda^2 - \alpha^2)^2 + \alpha^4\beta^2\}\right] \tag{3.2-9}$$

where

ν = dimensionless amplitude (maxima), $A/\sqrt{m_0}$
λ = dimensionless period, $T/\bar{T}_m$
m_0 = variance (area under wave spectrum)
$\bar{T}_m$ = average time interval of successive positive maxima = $4\pi(\sqrt{1-\varepsilon_2}/1 + \sqrt{1-\varepsilon_2}) \times \sqrt{m_0/m_2}$
$\alpha = (1/2)(1 + \sqrt{1-\varepsilon^2})$
$\beta = \varepsilon/\sqrt{1-\varepsilon^2}$
ε = band-width parameter of spectrum = $\sqrt{1 - m_2^2/m_0 m_4}$
m_2, m_4 = 2nd and 4th moment, respectively, of spectrum.

Equation (3.2-9) is modified to a dimensionless amplitude suitable for dealing with the breaking of equivalent Gaussian waves covered under a peak envelop process, and the average time interval of successive positive maxima is converted to an average zero-crossing period. Incorporating these changes of random variables in Equation (3.2-9), the following joint probability density function of wave amplitude and period applicable to non-Gaussian waves can be derived.

$$f(\xi,\tau)=\sqrt{\frac{2}{\pi}}\frac{\alpha^3\beta^2}{\varepsilon^3}\frac{\xi^2}{\gamma^4\tau^5}\exp\left[-\frac{1}{2\varepsilon^2}\frac{\xi^2}{\gamma^4\tau^4}\{(\gamma^2\tau^2-\alpha^2)^2+\alpha^4\beta^2\}\right] \qquad (3.2\text{-}10)$$

where

$\xi=$ dimensionless amplitude of equivalent Gaussian wave, $A/\sqrt{m_e}$, m_e is given in Equation (3.2-8)
$\tau=$ dimensionless period, $T/\overline{T}_0$
$\overline{T}_0=$ average zero crossing period $=2\pi\sqrt{m_0/m_2}$
$\gamma=\alpha\beta/\varepsilon$.

In order to evaluate the probability of breaking of non-Gaussian waves by using Equation (3.2-10), the breaking criterion given in Equation (3.2-7) may be written as

$$\xi_*=A/\sqrt{m_e}=0.098\,\tau^2(\overline{T}_0^2/\sqrt{m_e})(\tanh kh)^2 \qquad (3.2\text{-}11)$$

The probability of occurrence of breaking for a specified water depth h can be evaluated from Equations (3.2-10) and (3.2-11) by

$$\Pr\{\text{Breaking}\}=\int_0^\infty\int_{\xi_*}^\infty f(\xi,\tau)\mathrm{d}\xi\,\mathrm{d}\tau \qquad (3.2\text{-}12)$$

Care must be taken in evaluating the probability of occurrence of wave breaking in finite water depth by Equation (3.2-12), i.e. this equation includes waves which exceed the breaking criterion irrespective of their magnitude over the limiting height. Thus, computed results may include the breaking of small waves whose intensity is of such minor grade that the energy associated with breaking is almost nil. It may be well to explain this issue in detail using a practical example in the following.

Figure 3.7 shows an example of the computation of probability of breaking for hurricane KATE at a water depth 8.0 m with the joint probability density function given in Equation (3.2-10). The dashed line in the figure is the breaking criterion ξ_* given in Equation (3.2-11). As seen in the figure, the criterion of breaking line sharply increases with increase in τ (dimensionless period) up to a certain τ-value and then flattens out. This trend is a feature commonly observed in evaluating wave breaking in finite water depth. On the contrary, for wave breaking in deep water the criterion line uniformly increases with increase in τ. The probability of wave breaking is the volume of the joint probability density function above the ξ_*-line; the value is 0.202 in this example.

It is observed in Figure 3.7 that the volume of the joint probability density function above the breaking criterion line for small τ-values, say τ approximately less than 0.75 (periods shorter than 5.27 s), is fairly large. However, it will be shown later in Figure 3.8 that there is no appreciable energy loss of wave breaking in the frequency domain ω greater than 1.20 (periods shorter than 5.24 s). This implies that waves with a period shorter than 5.24 s (or τ smaller than 0.745 in Figure 3.7) break, but the severity of breaking is insignificant (may be the spilling status) in the spectrum. Hence, the probability of occurrence of wave

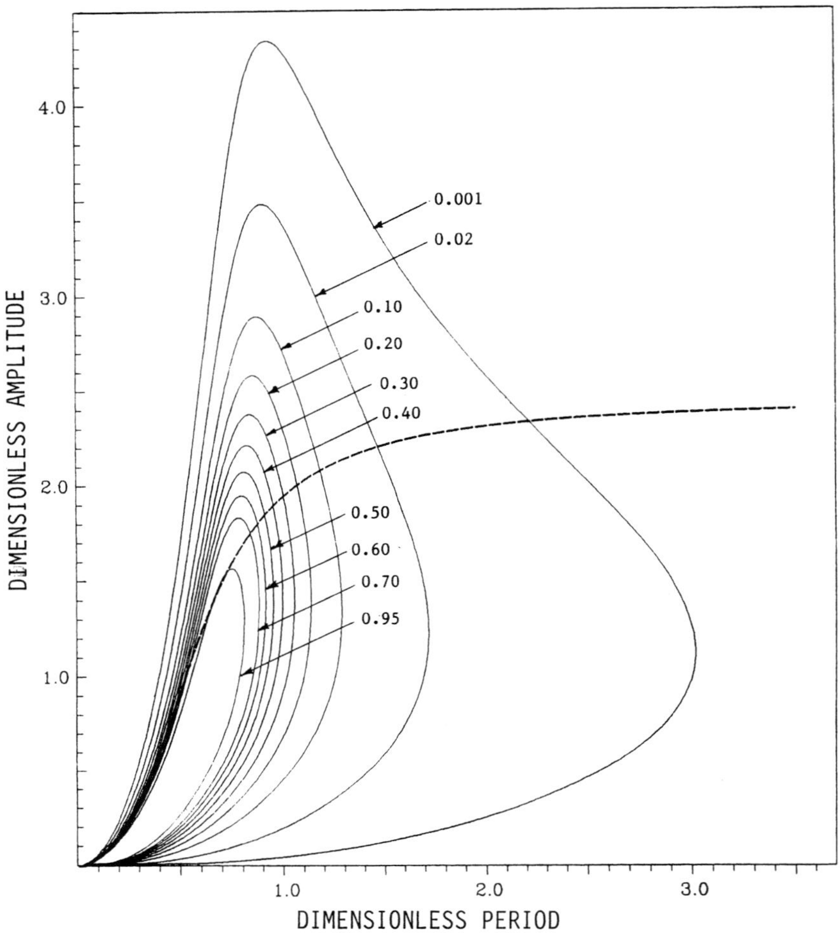

Figure 3.7 – Joint probability density function of wave amplitude and period including breaking criterion for non-Gaussian waves (Water depth 8.0 m).

breaking should be evaluated in the (ξ_e, τ) domain for τ greater than 0.745 in the example shown in Figure 3.7. The results of computations thus evaluated yield a probability of breaking of 0.110 in this example.

Thus, in summary, for a given water depth, it is necessary to evaluate the energy loss and obtain the frequency range where the energy loss is observed in the wave spectrum prior to evaluating the probability of wave breaking. Upon clarifying the frequency range, the probability of breaking can be evaluated by integrating the joint probability distribution of wave amplitude and period over the appropriate domain.

3.2.3 Energy loss associated with breaking

A method for evaluating the amount of energy loss in the wave spectrum due to wave breaking is developed by Longuet-Higgins (1969) in which he considers the breaking wave

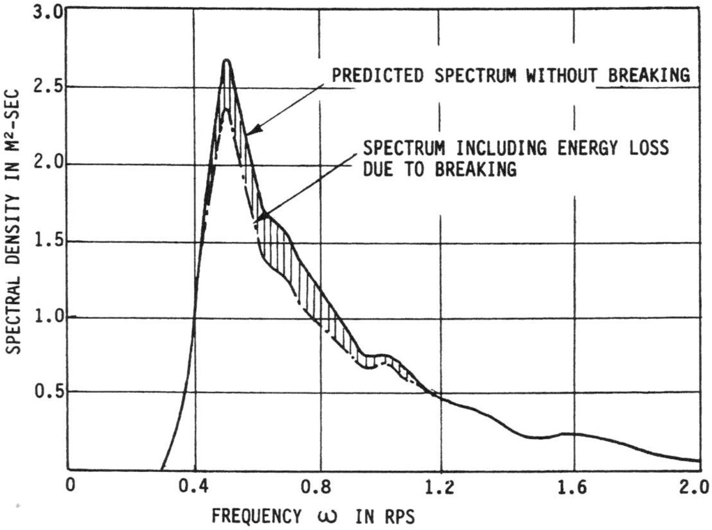

Figure 3.8 – Energy loss of hurricane KATE wave spectrum at the breaking point.

to reduce its height to the breaking limit height. In other words, for a given frequency of the wave spectrum, the energy loss associated with breaking can be evaluated (ignoring ρg) as follows

$$\Delta S(\omega) = \int_{A_*}^{\infty} \tfrac{1}{2}(A^2 - A_*^2)\, f(A|T)\mathrm{d}A, \qquad \omega = 2\pi/T \tag{3.2-13}$$

where A_* is the lower limit of amplitude A given in Equation (3.2-7) and $f(A|T)$ is the conditional probability density of amplitude for a given period.

Then, the ratio of loss of energy for a given frequency becomes,

$$\mu = \frac{\displaystyle\int_{A_*}^{\infty} \tfrac{1}{2}(A^2 - A_*^2)\, f(A|T)\mathrm{d}A}{\displaystyle\int_{0}^{\infty} \tfrac{1}{2}A^2\, f(A|T)\mathrm{d}A} \tag{3.2-14}$$

Since we are considering hurricane-generated non-Gaussian waves in finite water depth, we may apply Longuet-Higgins' concept to the equivalent Gaussian waves discussed in the previous section. For this, we may write Equation (3.2-14) in dimensionless form as

$$\mu = \frac{\displaystyle\int_{\xi_*}^{\infty} (\xi^2 - \xi_*^2)\, f(\xi|\tau)\mathrm{d}\xi}{\displaystyle\int_{0}^{\infty} \xi^2\, f(\xi|\tau)\mathrm{d}\xi} \tag{3.2-15}$$

where $f(\xi|\tau)$ is the dimensionless conditional probability density of wave amplitude for a given period.

Equation (3.2-15) can be presented in closed form, i.e. by integrating Equation (3.2-10) with respect to ξ, we have

$$f(\tau) = \frac{\alpha^3 \beta^2 \gamma \tau}{\{(\gamma^2 \tau^2 - \alpha^2)^2 + \alpha^4 \beta^2\}^{3/2}} \tag{3.2-16}$$

Then, the conditional probability density function $f(\xi|\tau)$ can be derived as

$$f(\xi|\tau) = f(\xi, \tau)/f(\tau)$$

$$= \sqrt{\frac{2}{\pi}} \frac{\{(\gamma\tau^2)^2 + \alpha^4\beta^2\}^{3/2}}{\varepsilon^3} \frac{\xi^2}{(\gamma\tau)^6}$$

$$\times \exp\left[-\frac{1}{2\varepsilon^2} \frac{\xi^2}{(\gamma\tau)^4} \{(\gamma^2\tau^2 - \alpha^2)^2 + \alpha^4\beta^2\} \right] \tag{3.2-17}$$

From Equations (3.2-15) and (3.2-17) and by letting $\tau = 2\pi/\omega\bar{T}_0$, the energy loss ratio can be evaluated as a function of frequency ω, i.e.

$$\mu(\omega) = \frac{2}{\sqrt{\pi}} \sqrt{\eta}\, \xi_* \,\mathrm{e}^{-\eta \xi_*^2} + 2\left(1 - \frac{2}{3}\eta \xi_*^2\right) \Phi(-\sqrt{2\eta}\, \xi_*) \tag{3.2-18}$$

where

$$\eta = \frac{1}{2\varepsilon^2} \frac{(\gamma^2 \tau^2 - \alpha^2)^2 + \alpha^4 \beta^2}{(\gamma\tau)^4}$$

and $\Phi(\ )$ is the cumulative distribution function of the standardized normal distribution.

Note that both η and ξ_* are functions of the variance m_e and dimensionless period of the equivalent Gaussian waves which are in turn a function of water depth.

The ratio of the total lost energy in the spectrum to the energy of the original spectrum before breaking, defined as the total energy loss ratio, can be evaluated by

$$\text{Total energy loss ratio} = \frac{\int_0^\infty \mu(\omega) S(\omega) \mathrm{d}\omega}{\int_0^\infty S(\omega) \mathrm{d}\omega} \tag{3.2-19}$$

The shape of the wave spectrum after energy is lost due to breaking, denoted by $S_*(\omega)$, can be evaluated by

$$S_*(\omega) = \{1 - \mu(\omega)\} S(\omega) \tag{3.2-20}$$

Figure 3.8 shows the estimated energy loss in the wave spectrum due to wave breaking computed for a water depth of 8.0 m during hurricane KATE. As seen in the figure, there is no appreciable energy loss associated with wave breaking at frequencies greater than 1.2 rps, which implies that waves with periods shorter than 5.24 s do not have any measurable energy loss (even though minor wave breaking may occur) in the spectrum, and thereby this period range should not be considered in evaluating the probability of wave breaking.

An interesting and important subject on wave breaking is the identification of the location where breaking takes place, commonly called the breaking point. The location depends on sea severity and water depth, and it can be evaluated for regular waves, but the location is not a fixed point and is unpredictable in random seas. It may be possible, however, to find a certain water depth where the frequency of occurrence of breaking as well as the energy loss associated with breaking both become sufficiently large so that breaking will affect the sea severity in the surf zone. This location may be defined as the breaking point in random seas.

Ochi and Malakar (2001) analyze wave data in storm seas obtained by the Coastal Engineering Research Center during the ARSLOE Project and find that a specific depth where the probability of breaking as well as the spectrum energy loss ratio both are at least 0.1 can be considered as the breaking point for a given sea severity. It may be well to give a more detailed explanation on this subject as follows.

Wave spectra at various water depths at Duck, North Carolina, are estimated by applying Kitaigorodskii's transformation factor to the measured deep water spectrum, and the probability of wave breaking as well as energy loss ratio associated with breaking are evaluated at each location. The results of the computations are plotted in Figure 3.9 as a function of water depth. As seen, the general trend of the total energy loss ratio in the wave spectrum is similar to that observed in the probability of occurrence of wave breaking, and the water depth where the probability of breaking and the total energy loss ratio are both at least 0.1 is 4.7 m in this example.

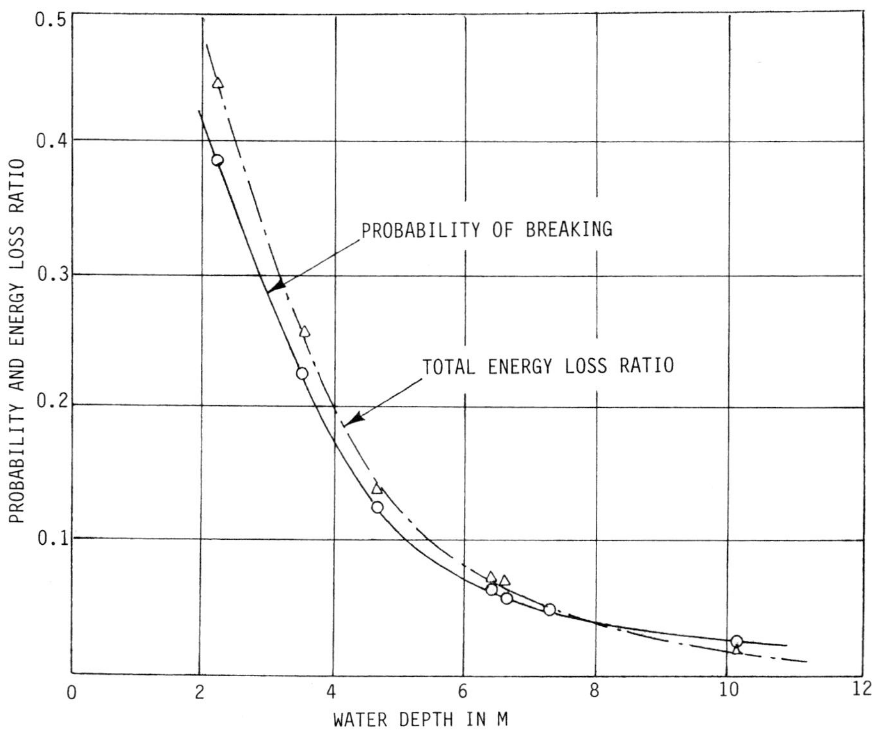

Figure 3.9 – Probability of wave breaking and total energy loss ratio due to breaking computed at various water depths off Duck, North Carolina (Data by Coastal Engineering Research Center).

It is found based on further computations and from comparison with the measured spectra that

(i) Measured spectra at water depths deeper than 4.7 m agree well with the estimated wave spectra including energy loss due to breaking.
(ii) Measured spectra at water depths less than 4.7 m agree reasonably well with the spectra estimated by applying Kitaigorodskii's transformation factor to the wave spectrum at a water depth of 4.7 m including the energy loss due to breaking. This will be presented in Section 3.3.

Thus, wave spectra in the surf zone depend on the wave spectrum including the effect of breaking (which may be called the broken wave spectrum) at a water depth of 4.7 m and thereby the water depth of 4.7 m is considered as the breaking point for this example.

Following the discussion presented above, the frequency of occurrence and energy loss ratio associated with breaking are evaluated for hurricane KATE spectra at various water depths and the results are shown in Figure 3.10. The spectra at various water depths are estimated by applying the transformation factor to the measured spectrum in deep water. From this figure, the water depth where the probability of breaking as well as the energy loss ratio both attain 0.1 is estimated as 8.3 m which is considered as the breaking point for hurricane KATE.

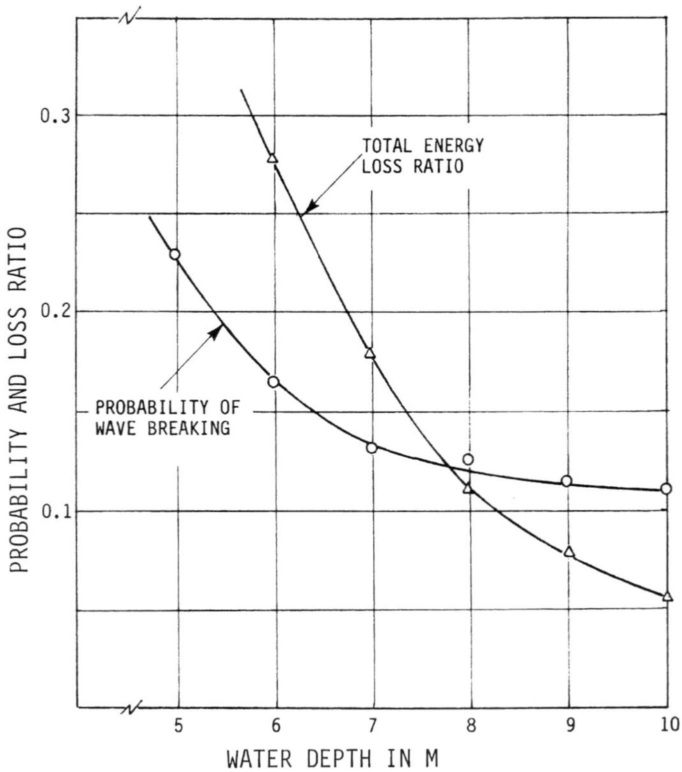

Figure 3.10 – Probability of wave breaking and total energy loss ratio due to breaking at various water depths computed in hurricane KATE.

3.3 Nearshore wave spectra

Evaluation of the transformation of wave spectra in the surf zone is extremely complicated since wave shoaling, breaking, refraction, wave-bottom interaction, etc. must be considered. Among these, the prime physical process to be considered for estimating hurricane-generated seas is the energy loss (dissipation) associated with breaking waves. In the surf zone, waves broken at the breaking point are traveling and the rate of depth variation is much more rapid than on the continental shelf.

A considerable number of studies has been carried out for evaluating sea severity in the surf zone through different concepts. Among these studies, an interesting and expedient approach is the application of the Boussinesq equations which are suitable for waves in shallow water incorporated with the effect of breaking waves by including dispersion of energy estimated based on the concepts of eddy viscosity, surface roller, or bore. Many publications on this subject have been presented in the last decade. These include Zelt (1991), Karambas and Koutitas (1992), Schaffer, Madsen, and Deigaard (1993), Nwogu (1993), Battjes, Eldeberky, and Won (1993), Eldeberky and Battjes (1996), Wei *et al* (1995, 1999), Raubenheimer *et al* (1996), Chen *et al* (1997, 2000), Kennedy *et al* (2000) and a series of Madsen and co-authors' work (1991, 1992, 1993, 1994, 1997, and 1998), etc. In these studies, comparisons between computed spectra and spectra obtained from experimental or field date in the surf zone (Eldeberky and Battjes, Chen and Guza, Raubenheimer and Guza, for example) and comparisons between computed and measured time histories of wave profiles (Eldeberky and Battjes, for example) all show very reasonable agreement, though all comparisons are made in either moderate or mild sea severity.

The majority of the studies on dissipations due to breaking waves incorporated with Boussinesq equation are for mild breaking; namely, spilling breakers. Studies made by Nwogu, Madsen, Wei, etc. however, deal with non-linear, non-Gaussian extension of Boussinesq equations and this is a promising approach for future estimation of severe sea states associated with hurricanes in the time domain.

There exist very little nearshore wave data measured at the time of hurricane landing (except that obtained in seas of mild severity) because either no measurement device was available at the site or the device was damaged by the severity of the sea. Hence, it is difficult to clarify the general characteristics and severity of wave spectra at the time of hurricane landing. One way to estimate wave spectra, however, is to derive the general characteristics of wave spectra as they are transformed in the surf zone from analysis of data obtained during the ARSLOE Project. We may assume that the transformation characteristics found from analysis of wave spectra in severe seas in the ARSLOE Project may also be applicable to estimate nearshore wave spectra at the time of hurricane landing.

We may first examine how the shape of nearshore wave spectra is transformed as the wave approaches the shoreline (Ochi and Malakar, 2001). Figure 3.11 shows the three spectra obtained from simultaneously measured data in severe seas at water depths of 4.70, 3.40, and 2.32 m at Duck, North Carolina obtained by the Coastal Engineering Research Center. The water depth of 4.70 m is the breaking point in this sea state as discussed in connection with Figure 3.9. As seen in Figure 3.11, the modal frequencies of these three spectra are nearly the same and the reduction of energy in the wave spectra occurs in a consistent uniform manner with decrease in water depth; there is no indication of energy transfer from one frequency to another. This result agrees with the conclusion derived from an experimental study on the propagation of breaking waves in shallow water conducted by Battjes and Beji (1992). Furthermore, the magnitude of wave energy at high frequencies is almost constant. In other

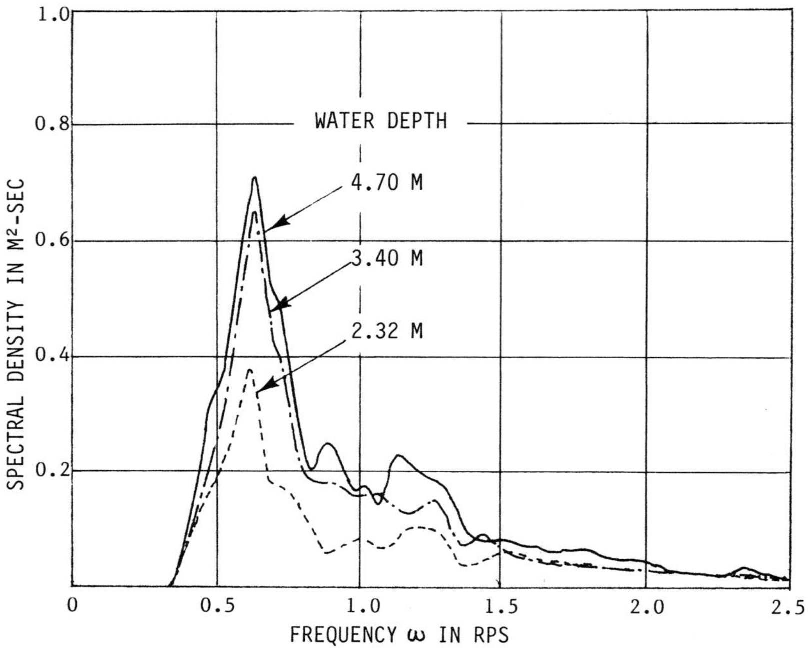

Figure 3.11 – Wave spectra obtained from measured data for three water depths in the surf zone during ARSLOE Project (Ochi and Malakar, 2001).

words, the general trend of energy dissipation observed in the figure indicates that the transformation of wave spectra in the surf zone after energy is lost at the breaking point and appears to uphold Kitaigorodskii's transformation factor. In order to confirm this, let us write the transformation factor for estimating the wave spectrum at a water depth h_2 from knowledge of a spectrum at a deeper water depth h_1 as follows.

$$S(\omega, h_2) = S(\omega, h_1)\, \Phi(\omega_h, h_1, h_2) \tag{3.3-1}$$

where

$$\Phi(\omega_h, h_1, h_2) = \left[\frac{k^{-3}(\omega, h_2) \dfrac{\partial}{\partial \omega} k(\omega, h_2)}{k^{-3}(\omega, h_1) \dfrac{\partial}{\partial \omega} k(\omega, h_1)} \right]_{\omega = \omega_h \sqrt{g/h}}, \qquad h_2 < h_1$$

Here, we choose the spectrum $S(\omega, h_1)$ to be that at the breaking point including the energy loss at that location; the broken wave spectrum.

We estimate the wave spectra for 3.40 and 2.32 m water depths by Equation (3.3-1) from knowledge of the broken wave spectrum obtained from data measured at a water depth of 4.70 m. Comparisons between computed and measured wave spectra for these depths are shown in Figure 3.12. As seen, the overall agreement is satisfactory; the difference in magnitude of total energy (the area under the spectrum) is 4.0 and 2.2% for water depths of 3.40 and 2.32 m, respectively. Thus, it is found that wave spectra in the surf zone may be

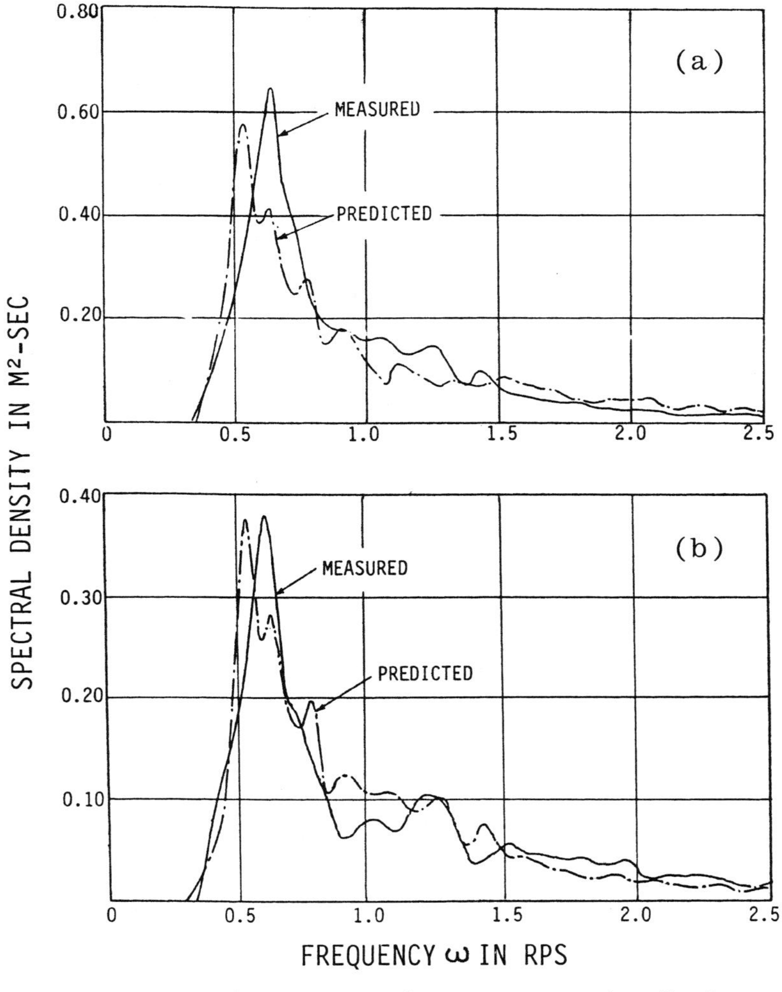

Figure 3.12 – Comparison between measured wave spectrum and predicted spectrum by applying Kitaigorodskii's transformation factor to broken wave spectrum at the breaking point: (a) Water depth 3.40 m; (b) water depth 2.32 m (Ochi and Malakar, 2001).

estimated by applying Kitaigorodskii's transformation factor to the broken wave spectra at the breaking point. This implies that the broken waves are propagating toward the shoreline in the surf zone, and that Kitaigorodskii's transformation factor accounts for the energy dissipation of the broken waves due to depth reduction.

The question arises as to the possible energy loss associated with additional breaking which might occur as the broken waves travel from the breaking point and approach the shoreline. In order to examine this possibility, computation of the wave spectrum at 2.32 m water depth is made on the traveling broken waves including additional breaking at 3.40 m water depth. Comparison of the results with the measured spectrum at 2.32 m shows that the computed spectrum is much smaller than the measured one; hence, it may be concluded that additional wave breaking consideration to the traveling broken waves evaluated at the breaking point is not necessary in the surf zone (Ochi and Malakar, 2001).

Based on the findings discussed in the above paragraphs, hurricane KATE wave spectra in the surf zone are estimated by applying the transformation factor to the broken wave spectrum at the breaking point shown in Figure 3.8, and the computed results are shown in Figure 3.13.

Figure 3.14 shows the wave energy (the area under wave spectrum) of hurricane KATE during the course of travel from deep to shallow water including the loss of energy associated with wave breaking at the breaking point (8.30 m) and shoaling in the surf zone.

It is of interest to note that, in general, one wave spectrum having larger energy than another does not necessarily maintain its status in the surf zone. As mentioned in Section 3.1,

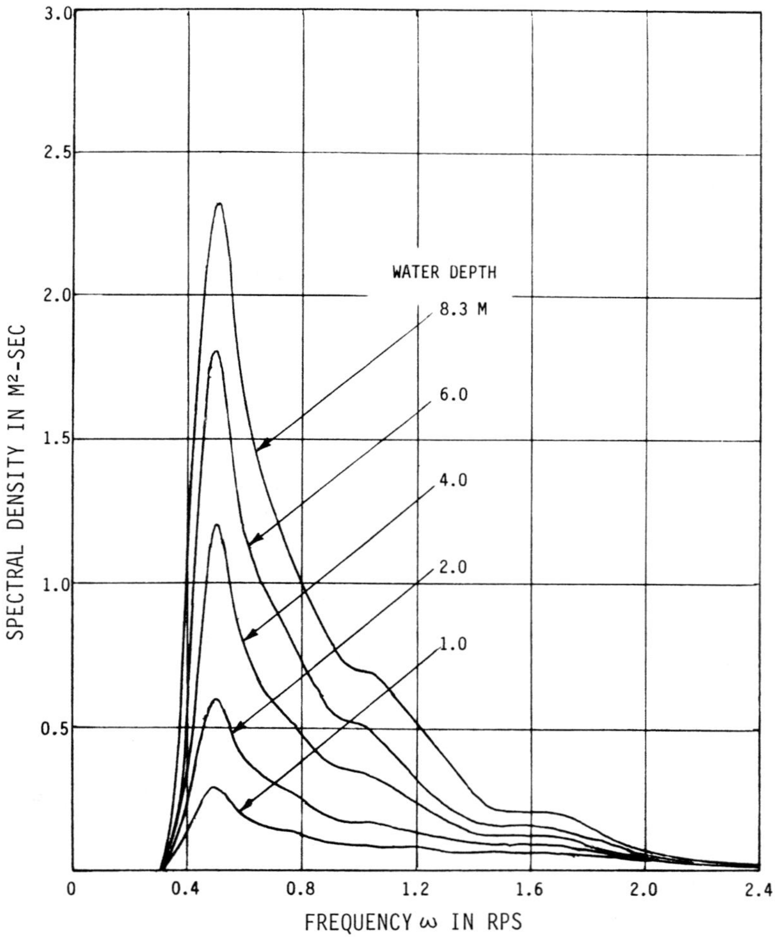

Figure 3.13 – Computed wave spectrum of hurricane KATE for various water depths in the surf zone.

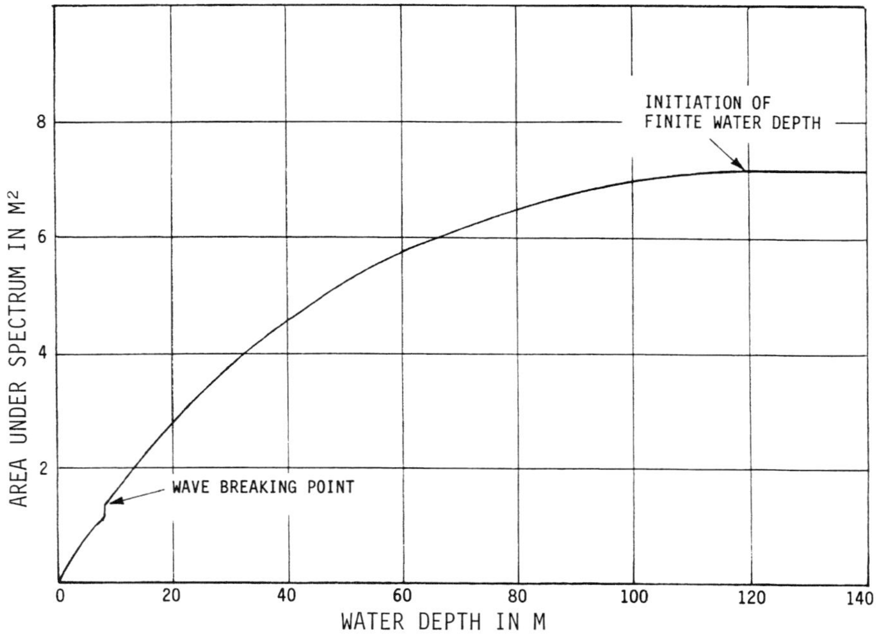

Figure 3.14 – Wave energy computed in hurricane KATE during the course of travel from deep to shallow water.

Kitaigorodskii's transformation factor is a function of frequency; a wave spectrum in deep water having a lower modal frequency loses more energy in finite water depth than one having a higher modal frequency. As an example, the sea severity measured in deep water generated by hurricane GLORIA is much greater than that in hurricane ELOISE; significant wave heights are 14.3 and 8.1 m, respectively. However, the energy of the computed wave spectrum of the former is less than that of the latter in the surf zone; the spectral areas are 0.268 m^2 for GLORIA *versus* 0.312 m^2 for ELOISE at a water depth of 3 m (Ochi, Malakar, and McClellan, 2000). This is not a contradictory result. The modal frequency for the deep water spectrum of hurricane GLORIA is 0.40 rps as compared with 0.55 rps for hurricane ELOISE. Hence, GLORIA lost more energy during its course of travel from deep to shallow water than hurricane ELOISE.

4

Sea severity and wave characteristics

In Chapters 2 and 3, the severity of hurricane-generated seas was presented in terms of the spectrum; namely, from the stochastic process, frequency point of view. In this chapter, the severity of sea condition in the time domain will be discussed including extreme wave height expected during passage of a hurricane.

4.1 Waves in deep water

Sea conditions observed in hurricane-generated seas are undoubtedly violent and hence wave characteristics are often considered to be totally non-linear. At present, we do not have sufficient information to verify the non-linear characteristics from data obtained in deep water during a hurricane. However, a wave record obtained during an extremely severe storm (significant wave height 16 m) in the North Atlantic, shown in Figure 4.1, indicates that the waves are Gaussian with only minor non-narrow-banded properties. Non-linearity of the waves shown in the figure seems to be minor, since the skewness is very small, 0.18 for this example. Hence, we may assume, in general, that waves of hurricane-generated seas in deep water can be considered as a Gaussian random process. Waves, however, may be narrow-band or non-narrow-band depending on the wind speed, hurricane speed of advance as well as the sea state prior to hurricane arrival.

If hurricane-generated waves in deep water are narrow-band Gaussian waves, then wave heights obey the Rayleigh probability distribution. Some claim that the probability distribution of wave height observed during a hurricane does not follow the Rayleigh probability law, or that the magnitude of positive amplitudes is greater than that of troughs. However, records indicating these properties are usually measured in finite water depths. There are some examples of wave records obtained during severe storms in deep water in which the magnitude of peaks is slightly greater than that of troughs; however, the skewness of these records is very small (less than 0.2) and thereby these waves can be considered essentially a Gaussian random process.

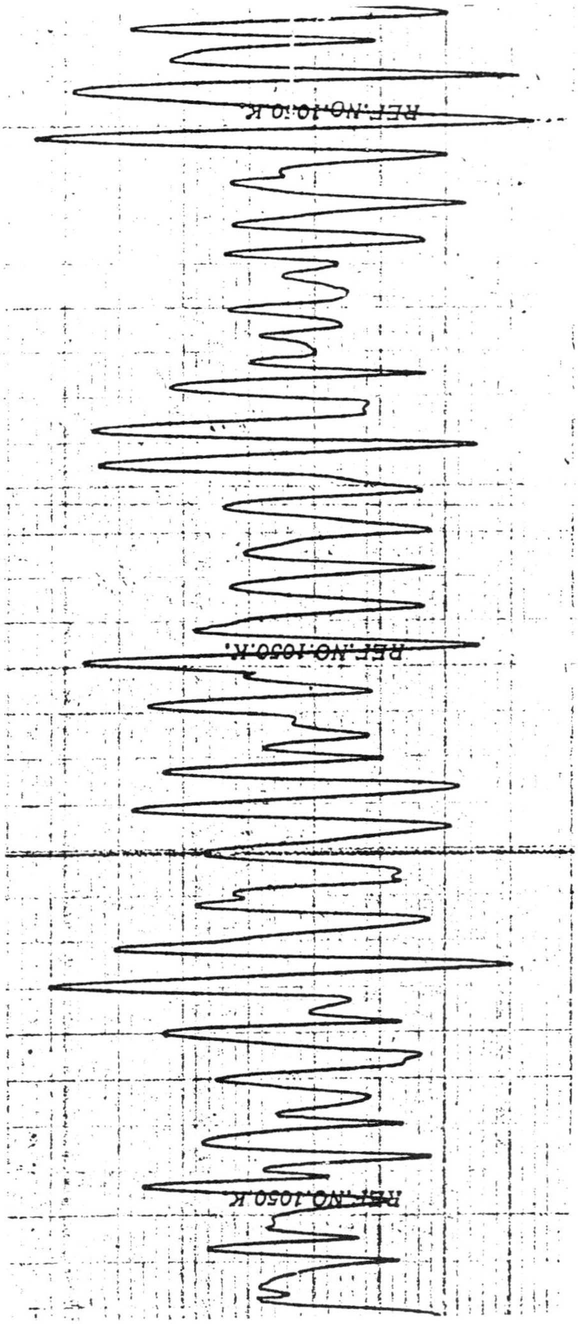

Figure 4.1 – Portion of time history of wave profile in very severe sea state in North Atlantic measured by a weather ship.

Thus, wave characteristics of hurricane-generated seas in deep water may be evaluated by the following formulae which are developed based on the narrow-band Gaussian wave concept.

- Significant wave height

$$H_s = 4\sqrt{m_0} \tag{4.1-1}$$

- Average wave height

$$\bar{H} = 2.52\sqrt{m_0} \tag{4.1-2}$$

- Average zero-crossing period

$$\bar{T} = 2\pi\sqrt{m_0/m_2} \tag{4.1-3}$$

- Band-width parameter

$$\varepsilon = \sqrt{1 - m_2^2/m_0 m_4} \tag{4.1-4}$$

- Average number of waves per unit time

$$n = \frac{1}{2\pi}\sqrt{m_2/m_0} \tag{4.1-5}$$

- Probable extreme wave height expected to occur in T-hours

$$\bar{H}_{\text{ext}} = 2\sqrt{2\ln\left(\frac{(60)^2 T}{2\pi}\sqrt{\frac{m_2}{m_0}}\right)}\sqrt{m_0} \tag{4.1-6}$$

- Extreme wave height for the design of marine systems operating in T-hours with the risk parameter α

$$H_{\text{ext}} = 2\sqrt{2\ln\left(\frac{(60)^2 T}{2\pi\alpha}\sqrt{\frac{m_2}{m_0}}\right)}\sqrt{m_0} \tag{4.1-7}$$

where m_j is the jth moment of spectrum $(= \int_0^\infty \omega^j S(\omega)d\omega)$.

It is noted that the extreme value evaluated for a given time period (1 hour, for example) is not a function of the band-width parameter ε (Ochi, 1973, 1998b).

In regard to extreme wave height during a hurricane, Borgman (1973) states that the assembly of wave heights in the total storm are sample values taken from a statistical population whose characterizing parameters are changing with time. Sobey, Chandler, and Harper (1990) also state that the sea condition during a hurricane is not a stationary stochastic process. The sea severity certainly changes continuously in the case of a hurricane. Therefore, the question arises as to estimation of the extreme wave height in the entire duration of a hurricane. This is similar to estimating extreme loading experienced by a ship (or offshore structure) in her lifetime operation in various sea conditions. Results of computations on long-term loading (including all sea states expected in the ship's lifetime) have demonstrated that extreme loading occurs in the piecewise steady state severest sea condition experienced by the ship (or offshore structure). Thus, we may safely assume that the stochastic properties of appropriately subdivided partitions of a wave record are piecewise steady state, and thereupon it may be well to evaluate the extreme values every half hour and examine the rate of change of extreme wave height depending on the varying situation of sea state and the advancing speed of the hurricane.

As an example, Figure 4.2 shows the results of computed significant wave height, probable and design extreme heights at every half-hour interval before and after the maximum sea state was reached when hurricane KATE passed nearby the NOAA Buoy 42003. These values are computed from measured wave spectra. Interim wave spectra at half-hour intervals are estimated by interpolating wave spectral data obtained every hour. The extreme values plotted at -1 hour, for example, are those expected to occur from 45 to 75 min before the maximum sea state. As seen in the figure, the significant wave height increases by 3.5 m (50%) and the probable extreme height (observed in half hour) increases by 5.5 m (46%) during the 3 hours preceding the hurricane center's arrival at the site. Such a rapid increase in wave height cannot be observed in ordinary wind-generated seas.

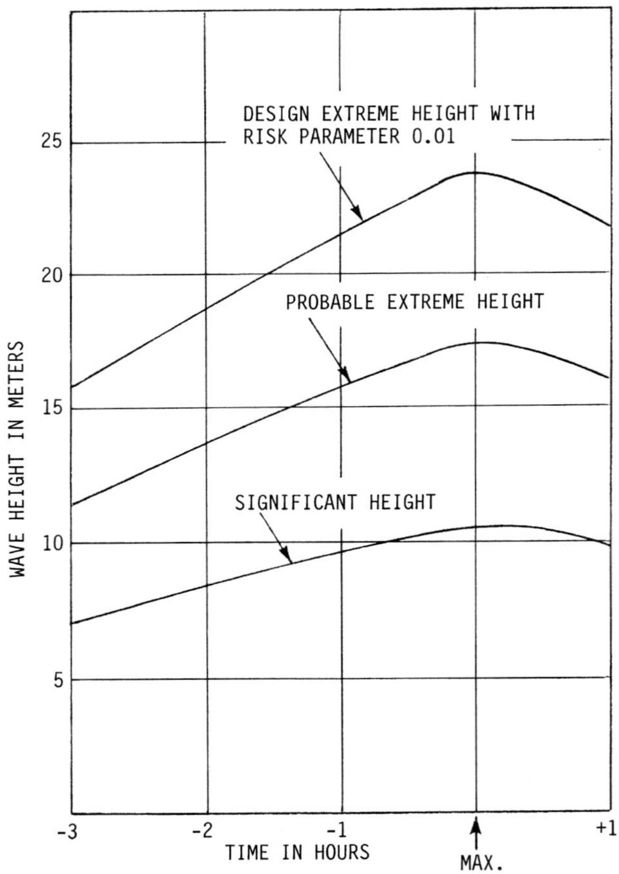

Figure 4.2 – Significant wave height and extreme height before and after the maximum sea state computed from wave spectra in hurricane KATE.

It is of interest to see the probability distribution of wave height constructed from data obtained when hurricane CAMILLE was about 38 km off station from the Ocean Data Gathering System (ODGS) in the Gulf of Mexico. The water depth at the recording site is finite (103.7 m in calm water) and the significant wave height at the time of recording was 12.3 m which is not the severest sea condition in the growing stage of hurricane CAMILLE.

Figure 4.3 shows a comparison between the Rayleigh probability function and a histogram of wave height presented in dimensionless form (wave height/$\sqrt{\text{variance}}$). Although some discrepancy between the Rayleigh distribution and the histogram is recognized in the figure, the data passes the χ^2-test for a level of significance of 0.05. Thus, we may assume that the hurricane CAMILLE waves at the time of this recording can be considered to be Gaussian and a narrow-band random process though the water depth is finite (103.7 m in calm water).

The extreme wave height recorded in a 30 min interval during hurricane CAMILLE was 22.6 m as compared with a probable extreme wave height of 19.8 m computed by Equation (4.1-6). The measured extreme wave height is greater than the estimated probable extreme wave height. This is not unusual, since it is known that the possibility of measured extreme wave height exceeding the estimated probable extreme height is theoretically $1 - e^{-1} = 0.632$. On the other hand, the design extreme wave height in 30 min with a risk parameter of 0.01 computed by Equation (4.1-7) is 27.2 m which is far greater than the measured largest wave height. This indicates that the extreme wave heights expected in hurricane-generated seas in deep water may be evaluated by applying the formulae given in Equations (4.1-6) and (4.1-7) developed for ordinary storm seas.

Wave characteristics of typical hurricane-generated seas in deep water evaluated from measured data are tabulated in Table 4.1.

It is of interest to estimate the severest sea condition expected to occur in hurricanes at a specific location over a long time period, 50 and 100 years for example. For this, information on the frequency as well as the severity of hurricanes at that location is a prerequisite, but in reality, the observed data are too small to provide sufficient information. In order to overcome this difficulty, Donoso, LeMehaute, and Long (1987) apply the Monte Carlo simulation

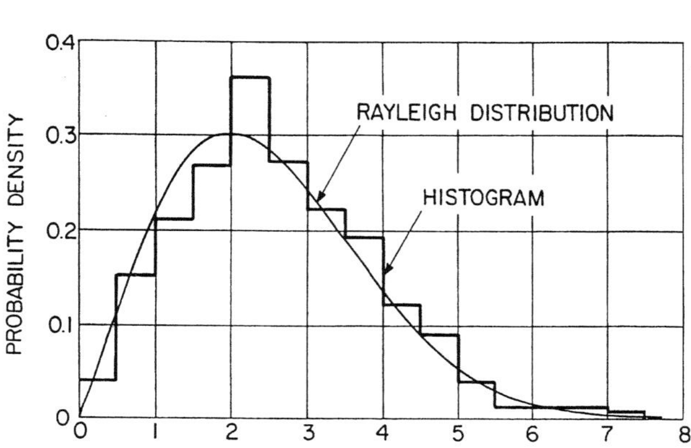

Figure 4.3 – Comparison between wave height histogram (dimensionless) obtained during hurricane CAMILLE and Rayleigh probability density function.

Measurements	BELLE	ELOISE	FREDERICK	GLORIA	KATE	GEORGE	CAMILLE ODGS platform[1]
	NOAA buoy						
	EB15	EB10	EB44	41002	42003	42040	
Location	33.0N 75.2W	27.5N 88.0W	26.0N 86.0W	32.3N 75.3W	25.9N 85.9W	29.2N 88.2W	29.0N 88.4W
Significant wave height (m)	7.26	8.07	8.78	14.27	10.70	11.00	13.20
Modal frequency (rps)	0.425	0.550	0.620	0.402	0.475	0.503	0.530
Average zero-crossing period (s)	9.35	8.66	8.56	11.80	10.00	10.02	8.92
Band-width parameter, ε	0.796	0.743	0.740	0.685	0.645	0.647	0.560
Probable extreme height in half hour (m)	11.8	14.3	14.5	22.6	17.3	17.7	22.2
Design extreme height in half hour with risk parameter 0.01 (m)	16.3	18.9	19.8	31.3	23.6	24.4	30.3

(1) Water depth 103.7 m in calm sea.

Table 4.1 *Wave characteristics of typical hurricane-generated seas in deep water*

method for generating sea state data, i.e. data on hurricane parameters such as central pressure, radius of maximum wind, forward speed, and the shortest distance to the hurricane in the vicinity of a specified site are accumulated and the probability distribution of each parameter is established. A synthetic storm is then generated by employing a hurricane model developed by Ross (1976, 1979) with a set of parameters drawn at random from their respective population. The synthetic population of sea state generated in this way is used to establish the probability distribution of sea state around the specified site. Independently, the Poisson process is considered for evaluating the probability of occurrence of hurricanes. By combining these two probability distributions, the return interval for any given sea state is estimated.

Although Donoso *et al*'s approach to estimate the extreme sea condition associated with hurricanes in the long-term is reasonable, great care has to be given to each probability distribution of the above mentioned four parameters and their statistical correlation from which the input to the simulation technique is generated. Some of the four parameters appear to be highly correlated; hence, a joint probability distribution may be necessary. This implies that, in practice, a great number of data are required in order to obtain reliable results by applying this simulation method for estimating hurricane associated extreme sea states in the long-term.

4.2 Waves in finite water depth

4.2.1 Transformation of Gaussian waves to non-Gaussian waves

The characteristics of waves of hurricane-generated seas in finite water depth are those observed in waves from the continental shelf to the breaking point at which non-linearity is introduced and thereby the stochastic properties are transformed from Gaussian random waves to non-Gaussian waves when the sea state becomes severe.

It was shown in Section 3.1 that the location where the effect of finite water depth initiates can be found by applying Kitaigorodskii's transformation factor. In principle, non-linear wave properties and non-Gaussian waves may initiate simultaneously at that location. However, in practice, deviation of the statistical distribution of the wave profile from the normal probability law is extremely small at the initial stage of the water depth becoming finite, i.e. histograms of wave profiles at the initial stage of finite depth may indicate a small deviation from the normal probability distribution, but the data usually passes the χ^2-test with a level of significance of 5%. When the χ^2-test fails at a certain water depth in a given sea state, then we must consider non-Gaussian waves (or non-linearity) for the design of marine systems.

Let us examine the limiting water depth for a given sea severity where the transformation from Gaussian to non-Gaussian waves must be considered. Ochi and Wang (1984) analyze several hundred wave records obtained by the Coastal Engineering Research Center during the ARSLOE Project and evaluate the skewness and kurtosis, denoted by λ_3 and λ_4, respectively, of each record. The results of analysis of the ARSLOE field data show that a functional relationship exists between the two parameters λ_3 and λ_4. The relationship agrees very well with that obtained from experiments in the laboratory by Kobayashi *et al* (1998).

In the analysis of ARSLOE data, histograms of wave profiles are constructed for each record and compared with the Gram-Charlier series Type A probability density function as well as the normal probability density function. The results of comparison show that the probability distribution of wave profiles with λ_3 less than 0.2 are very close to the Gaussian distribution. Almost all data having skewness of 0.2 pass the χ^2-test at the 5% level of significance. Hence, it may be concluded that skewness of $\lambda_3 = 0.2$ is the limiting value for classifying whether or not waves are considered to be a non-Gaussian random process.

Based on this finding, wave data in the ARSLOE Project having $\lambda_3 = 0.2$ (or very close thereto) in various sea severities and water depths are accumulated, and the relationship between significant wave height and water depth is obtained; the results are shown by the dotted line $\lambda_3 = 0.2$ in Figure 4.4. This, in turn, indicates the limiting water depth below which waves are considered to be non-Gaussian for a given sea state.

On the other hand, Robillard and Ochi (1996) reanalyze the same ARSLOE wave data expressing the probability distribution of the wave profile by the non-Gaussian probability distribution function presented in the Appendix. One of the three parameters of the non-Gaussian distribution, denoted by a represents the non-linearity; $a = 0$ for Gaussian random waves. Wave data having the computed parameter a of the distribution being zero or very close thereto are accumulated (a total of 11 data), and the significant wave height and water depth of these data are plotted as the open circles in Figure 4.4.

Additionally, all wave data are classified into Gaussian or non-Gaussian groups by comparing the histogram and the normal probability distribution of the wave profiles, and presented as a function of σ/h, where σ is the rms-value and h, the water depth. The χ^2-test

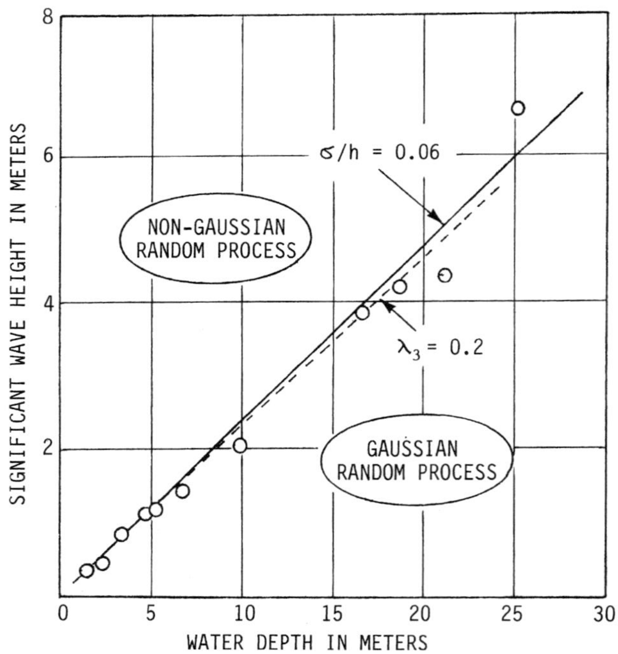

Figure 4.4 – Limiting water depth below which waves are considered to be non-Gaussian as a function of significant wave height (Ochi and Wang, 1984; Robillard and Ochi, 1996).

with level of significance of 5% is used for decision making in classifying histograms which deviate slightly from the normal probability distribution. It is found that all histograms of wave profiles having the ratio σ/h less than 0.06 obey the normal probability distribution. Assuming that the significant wave height, H_s is equal to 4σ, this results in the ratio $h/H_s >$ 4.17 obeying the normal probability distribution. Hence, $h = 4.17H_s$ is the limiting water depth for a given sea state below which the waves are considered to be non-Gaussian. This result is also included in Figure 4.4 as the solid line $\sigma/h = 0.06$. It should be noted that the assumption of $H_s = 4\sigma$ slightly overestimates the significant height of non-Gaussian waves, but the approximation should not cause a serious problem. A detailed discussion on this subject will be given later.

In summary, as shown in Figure 4.4, the results obtained from three different approaches in the evaluation of the sea severity-water depth relationship agree very well. Therefore, it may be concluded that, for a given sea severity, the water depth which satisfies the relationship $\sigma/h = 0.06$ is the limiting water depth below which waves are considered to be a non-Gaussian random process.

Although the above-mentioned conclusion is derived from analysis of data obtained in seas of water depth 25 m or less, no other field data at greater depths are currently available from which we can estimate the water depth where Gaussian waves are transformed to non-Gaussian waves; hence, the assumption made at present is that the relationship $\sigma/h = 0.06$ holds in more deep water depths than 25 m.

Let us estimate the water depth at which the deep water Gaussian waves of hurricane KATE transform into non-Gaussian waves. It is known that the area under the deep water spectrum $(m_0)_\infty = 7.16\ \text{m}^2$, so from the relationship $\sigma/h = \sqrt{(m_0)_h}/h = 0.06$, we can derive the relationship $(m_0)_h/(m_0)_\infty = (h/44.6)^2$. Then, from Figure 3.5 we can obtain the water depth which satisfies this relationship as $h = 33.7$ m.

Summarizing the results of computations made on hurricane KATE we may say that the effect of finite water depth on the wave spectrum initiates at 120 m depth on the continental shelf (see Section 3.1), but wave properties can still be evaluated under the linear, Gaussian random wave concept. Non-linear, non-Gaussian wave properties, however, must be considered for water depths less than 33.7 m.

4.2.2 Evaluation of non-Gaussian waves

Upon finding the water depth below which waves are considered to be non-Gaussian using the criterion $\sigma/h \geq 0.06$ (where $\sigma =$ rms-value $= \sqrt{m_0}$, m_0 is the area under the spectrum, h the water depth), the probability density function representing the statistical characteristics of the wave profile can be determined from knowledge of the wave spectrum at a specified water depth. The non-Gaussian probability distribution is discussed in the Appendix in detail; hence, the procedure for deriving the probability density function is outlined here without detailed explanation.

The probability density function of non-Gaussian waves carries three parameters; a, μ_*, and σ_*. These parameters are a function of not only the specified water depth, h, during the storm, but also the calm water depth at that location, denoted by D_0. The depth h may be greater or less than D_0 depending on storm surge and/or tide. At the time when a hurricane comes close to the shoreline, however, the water depth h will most likely be greater than D_0. For a given wave spectrum at a specified water depth (h, D_0), let σ be the square-root of the area under the spectrum; then, the parameters of the non-Gaussian probability distribution can be evaluated as follows.

(i) Evaluate $a\sigma$ from Figure A.2 in the Appendix for a given σ/h and h/D_0, and determine the parameter a. For $\sigma/h > 0.12$, $a\sigma$ can be evaluated by Equation (A.1-9).

(ii) Evaluate μ_*/σ by

$$\mu_*/\sigma = -1.55(a\sigma)^{1.20} \qquad \text{(see Figure A.5)} \tag{4.2-1}$$

and determine the parameter μ_*.

(iii) Evaluate σ_*/σ by

$$\sigma_*/\sigma = \exp\{(a\sigma)^2\} \qquad \text{(see Figure A.6)} \tag{4.2-2}$$

and determine the parameter σ_*.

The probability density function of the wave profile (displacement from the mean value) of non-Gaussian waves is given by (Ochi and Ahn, 1994a,b)

$$f(x) = \frac{1}{\sqrt{2\pi}\sigma_*}\exp\left\{-\frac{1}{2(\gamma a\sigma_*)^2}(1-\gamma a\mu_* - \mathrm{e}^{-\gamma ax})^2 - \gamma ax\right\} \tag{4.2-3}$$

where $\gamma = 1.28$ for $x \geq 0$ and 3.0 for $x < 0$.

It can be proved that Equation (4.2-3) reduces to a normal probability density function with mean μ_* is zero and variance σ_*^2 if $a = 0$. An example of comparison between the probability density function given in Equation (4.2-3) and the histogram constructed from non-Gaussian wave data is shown in Figure A.1 in the Appendix.

Let us evaluate the probability distribution of the wave profiles of hurricane KATE at various water depths on the continental shelf by Equation (4.2-3) and examine how the skewness of the non-Gaussian probability distribution increases with decrease in water depth. For this, computations are made at various water depths ranging from 33.7 m (where the non-Gaussian wave property initiates) to the breaking point 8.3 m (before breaking takes place) and the results are shown in Figure 4.5. The input for evaluating the probability density function for

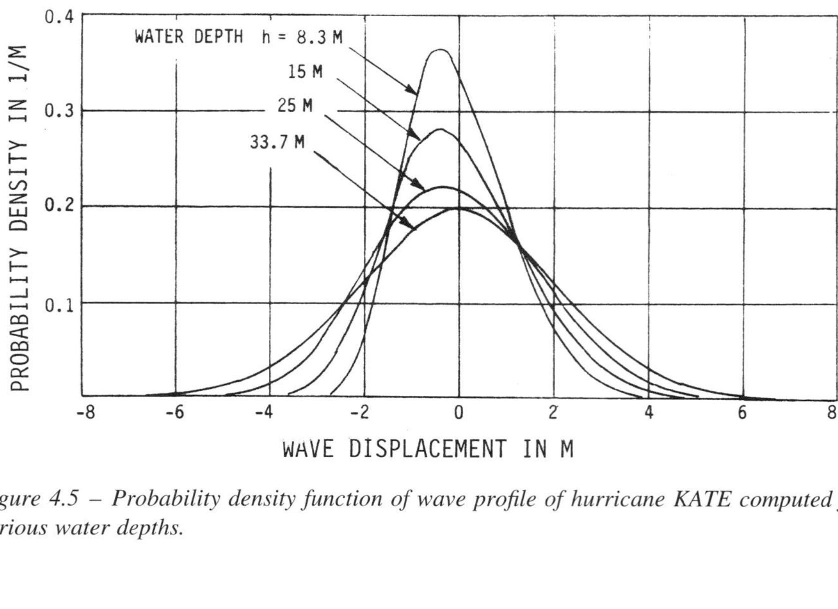

Figure 4.5 – Probability density function of wave profile of hurricane KATE computed for various water depths.

a specified water depth is the spectral density function (variance) at each depth computed from the deep water spectrum by applying the method discussed in Section 3.3. Computed variances were given earlier in Figure 3.14. As seen in Figure 4.5, the probability density function slowly deviates from the normal distribution with increasing skewness with decrease in water depth.

As mentioned earlier, non-Gaussian waves depend on sea severity and water depth, in general. Here, the water depth includes the variation of water level due to storm surge and tide, the effect of which are particularly pronounced in shallow water. The water depths used in the computations shown in Figure 4.5 are for $h/D_0 = 1$; namely, the water depth during the storm is the same as the calm water level.

In order to examine the effect of variations of water level associated with storm surge and/or tide on the probability density function of the wave profile, computations are made at a water depth of $h = 8.3$ m (before wave breaking) for $h/D_0 = 0.8$, 1.0, 1.2, and 1.4, and the results are shown in Figure 4.6. In these computations, the sea severity is maintained constant. Here, $h/D_0 = 0.8$ represents the situation where the water level of 10.4 m is reduced to 8.3 m in ebb tide, while $h/D_0 = 1.4$ represents the water level of 5.9 m being increased to 8.3 m due to storm surge and/or flood tide.

As seen in Figure 4.6, the probability density function of the wave profile for a specified water depth, h, is almost the same irrespective of the change of water level associated with storm surge and/or tide in this example. Thus, it appears that the location of

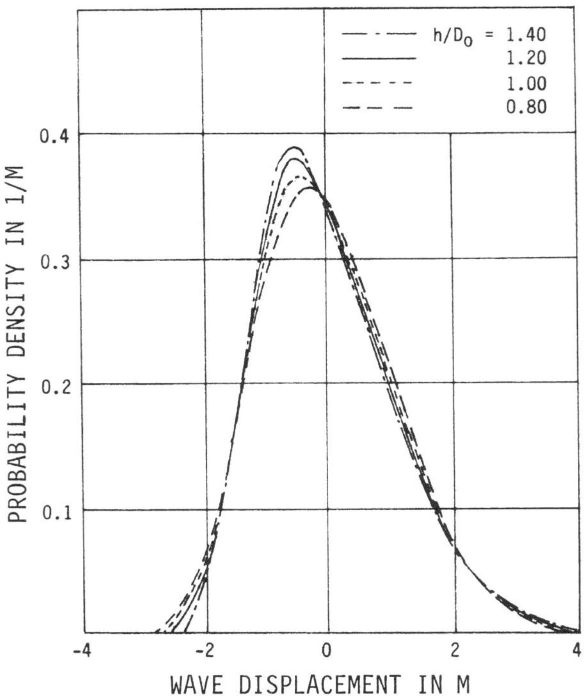

Figure 4.6 – Effect of h/D_0 (h is the specified water depth, and D_0 the calm water depth) on probability density function of wave displacement (hurricane KATE, $h = 8.3$ m).

the breaking point may be estimated without consideration of storm surge and/or tide for hurricanes of the same order of severity as KATE. These results, however, should not be confused with the effect of variation of water level for a specified calm water depth, D_0. The effect of h/D_0 on the probability distribution of the wave profile for a given D_0 depends on h/D_0, since the sea severity cannot be assumed constant.

4.2.3 Probability distribution of wave height

Information on the probability distribution of wave height in the area of the continental shelf is extremely important for the design and operation of offshore structures since the majority of the offshore structures are located on the continental shelf. It is highly desirable to evaluate the probability distribution of the peaks and troughs separately for waves on the continental shelf, since they may be non-Gaussian depending on sea severity and water depth whereby the difference between the shape of the peaks and troughs is significant. The derivation of the probability distribution of wave amplitudes and heights of non-Gaussian waves is given in the Appendix; hence, only the procedure to evaluate the probability density function is outlined below:

(i) *Probability distribution of peaks*

The probability distribution of the envelope of peaks (and troughs also) of non-Gaussian waves can be analytically derived as the sum of narrow-band Gaussian waves and a sine wave. In particular, if the skewness of the non-Gaussian waves is less than 1.2 (skewness of almost all waves is less than 1.2), then the probability density function can be approximately represented in the form of the Rayleigh probability distribution with sufficient accuracy (Ochi, 1998a). However, the parameter of the Rayleigh distribution becomes a function of three parameters, a, μ_*, and σ_*, for non-Gaussian waves as discussed in Section 4.2.2. The probability density function of peaks ξ of non-Gaussian waves is given by

$$f(x) = \frac{2\xi}{R_1} e^{-\xi^2/R_1}, \qquad 0 \le \xi \le \infty \tag{4.2-4}$$

where

$$R_1 = 2s_1^2/\{1 - (2c_1^2/5s_1^2)\}$$

$$s_1^2 = \sigma_1^2/(1 + \lambda_1 \mu_*)$$

$$c_1^2 = 2(\lambda_1 \sigma_1^2 - \mu_*)^2/(1 + \lambda_1 \mu_*)^2$$

$$\lambda_1 = 1.28a$$

$$\sigma_1^2 = \sigma_*^2\{1 + 2\sqrt{2/\pi}(\lambda\sigma_*) + 3(\lambda\sigma_*)^2\}$$

Thus, from knowledge of the three parameters of the probability distribution representing non-Gaussian waves, the parameter of the Rayleigh probability distribution R_1 can be determined. An example of comparison between the probability density function given in Equation (4.2-4) and the histogram constructed from non-Gaussian wave data is shown in Figure A.8 in the Appendix.

(ii) *Probability distribution of troughs*

The probability density function of the troughs η of non-Gaussian waves can be derived through the same concept as that for peaks, i.e. the density function can be presented in the form of the Rayleigh probability distribution as follows

$$f(\eta) = \frac{2\eta}{R_2} e^{-\eta^2/R_2}, \qquad 0 \leq \eta \leq \infty \tag{4.2-5}$$

where

$$R_2 = 2s_2^2/(1 - 2c_2^2/5s_2^2)$$

$$s_2^2 = \sigma_2^2/(1 + \lambda_2\mu_*)$$

$$c_2^2 = 2(\lambda_2\sigma_2^2 - \mu_*)^2/(1 + \lambda_2\mu_*)^2$$

$$\lambda_2 = 3a$$

$$\sigma_2^2 = 2\sigma^2 - \sigma_1^2$$

where σ^2 is the variance of the non-Gaussian waves.

Thus, again from knowledge of the three parameters of the probability distribution representing non-Gaussian waves, the parameter R_2 can be determined. An example of comparison between the probability density functions of peaks and troughs of non-Gaussian waves is shown in Figure A.9 in the Appendix.

(iii) *Probability distribution of wave height*

The probability density function of peak-to-trough excursions (wave height) ς can be derived as the probability distribution of the sum of two independent random variables ξ (peak) and η (trough). The derived probability density function becomes,

$$f(\varsigma) = \frac{2\varsigma}{(R_1 + R_2)^2}(R_1 e^{-\varsigma^2/R_1} + R_2 e^{-\varsigma^2/R_2}) + \frac{2\sqrt{\pi}}{R_1 + R_2}$$
$$\times e^{-\varsigma^2/(R_1+R_2)} \sqrt{\frac{R_1 R_2}{R_1 + R_2}} \left(\frac{2}{R_1 + R_2}\varsigma^2 - 1\right)$$
$$\times \left[\Phi\left(\sqrt{\frac{2R_2}{R_1(R_1 + R_2)}}\varsigma\right) - \Phi\left(-\sqrt{\frac{2R_1}{R_2(R_1 + R_2)}}\varsigma\right)\right] \tag{4.2-6}$$

where R_1 and R_2 are the Rayleigh distribution parameters given in Equations (4.2-4) and (4.2-5), respectively and $\Phi(\ )$ the cumulative probability distribution of the standardized normal probability function.

An example of comparison between the probability density function of non-Gaussian wave height and the histogram constructed from measured data is shown in Figure A.10 in Appendix.

(iv) *Significant wave height and extreme wave height*

The significant wave height of non-Gaussian waves can be analytically derived with some approximation, however, the accuracy is not always satisfactory. The numerically computed significant wave height, H_s, using the following procedure agrees well with that evaluated from observed data of non-Gaussian waves, i.e. first determine the value of s from

$$\int_{s}^{\infty} f(\varsigma)\mathrm{d}\varsigma = 1/3 \tag{4.2-7}$$

Then, significant wave height can be evaluated by

$$H_s = 3\int_{s}^{\infty} \varsigma f(\varsigma)\mathrm{d}\varsigma \tag{4.2-8}$$

To evaluate the extreme wave height, the number of waves per unit time, denoted by n, is first computed from the wave spectrum by Equation (4.1-5). The number of waves in half hour, denoted by $N_{1/2}$, thereby becomes $1800n$. The extreme values can then be obtained as follows:

(a) Probable extreme wave height in half hour, denoted by $\bar{y}$, can be obtained as the value of $\bar{y}$ which satisfies the following equation

$$\int_{\bar{y}}^{\infty} f(\varsigma)\mathrm{d}\varsigma = 1/N_{1/2} \tag{4.2-9}$$

(b) Design extreme wave height with the risk parameter α in half hour, denoted by $\hat{y}(\alpha)$, can be obtained as the value of $\hat{y}(\alpha)$ which satisfies the following equation

$$\int_{\hat{y}(\alpha)}^{\infty} f(\varsigma)\mathrm{d}\varsigma = \alpha/N_{1/2} \tag{4.2-10}$$

Figure 4.7 shows the probability density function of peak and trough amplitudes of hurricane KATE computed at four water depths on the continental shelf. The water depth of 8.3 m is the breaking point, but the distribution shown in the figure is that before the energy loss due to breaking occurred. The distribution including the energy loss due to breaking will be shown later. As seen in the figure, symmetry of the probability density function of peak and trough amplitudes associated with Gaussian waves observed at the water depth of 33.7 m gradually deforms with decrease in water depth. The asymmetric shape of the density function of peaks and troughs is substantial at the breaking point associated with hurricane-generated seas. This result suggests that the evaluation of breaking characteristics such as the frequency of occurrence of breaking should be made considering non-Gaussian waves.

The probability density function of wave height computed for hurricane KATE at four water depths on the continental shelf is shown in Figure 4.8. The density function for water depth 33.7 m is the Rayleigh probability distribution whose parameter is equal to eight times the area under the spectrum. The shape of the density function at water depths less than 33.7 m shown in the figure deviates significantly from that of the Rayleigh distribution, although the Rayleigh distribution is usually considered for predicting statistical properties of wave heights on the continental shelf.

In order to examine the difference between the probability distribution of wave height evaluated based on two concepts (non-Gaussian and Gaussian waves), Figure 4.9 is prepared. It shows the probability density function of non-Gaussian and Gaussian wave height computed at a water depth of 25 m. Here, the parameter of the probability distribution of Gaussian waves (Rayleigh distribution) is eight times the area under each spectrum. As seen, the Gaussian random process concept overestimates the probability of occurrence of large wave heights.

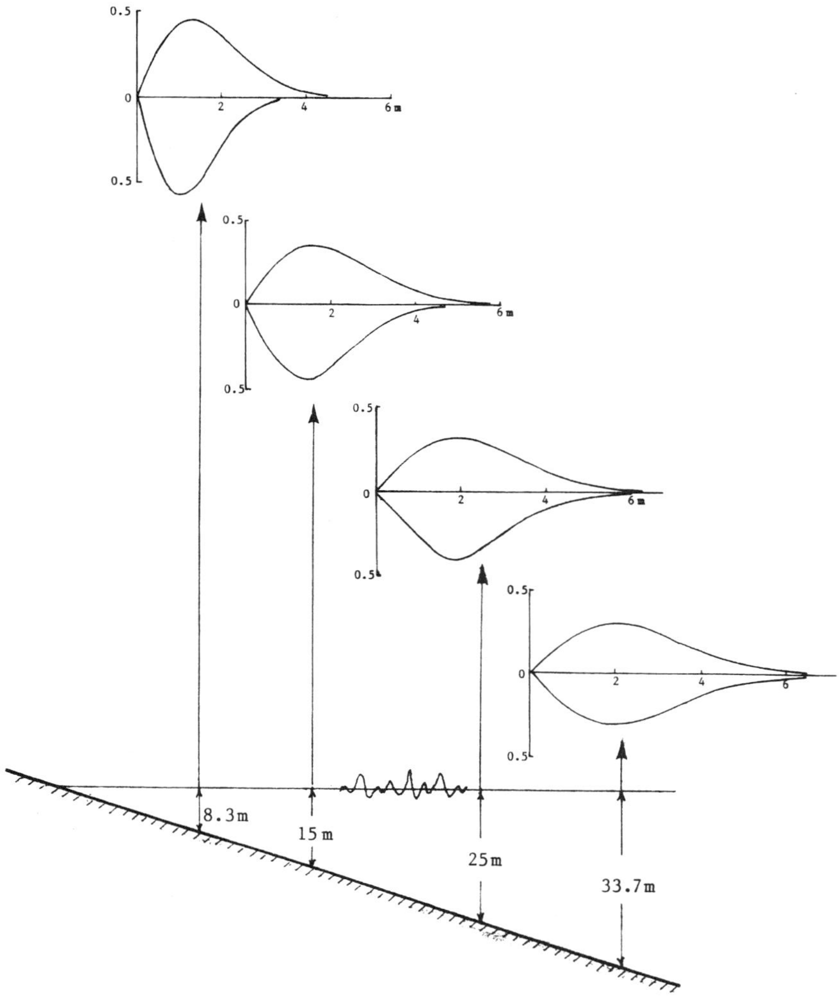

Figure 4.7 – Probability density function of wave peak and trough amplitudes computed for various water depths in hurricane KATE.

The cumulative distribution function of wave height, $F(H)$, may provide a more descriptive difference between the two concepts. Figure 4.10 shows $1 - F(H)$ which implies the probability of wave height exceeding a specified value, H, for hurricane KATE for a water depth of 25 m. The difference in the probabilities estimated by the two concepts is very large. For example, the probability of wave height exceeding 10 m is 0.003 for the non-Gaussian concept *versus* 0.024 for the Gaussian concept in this example.

The general trend of the difference between the non-Gaussian and Gaussian concepts observed in finite water depth is consistent; i.e. the Gaussian wave assumption (Rayleigh probability distribution) always overestimates the probability of occurrence of large wave heights in finite water depth. This general trend is also observed in the analysis of field data.

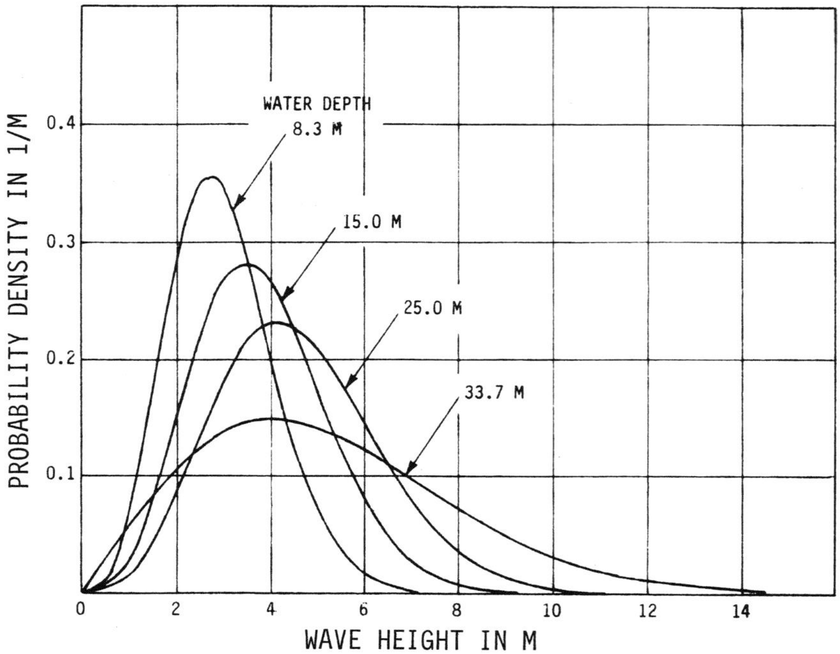

Figure 4.8 – Probability density function of wave height computed for various water depths in hurricane KATE.

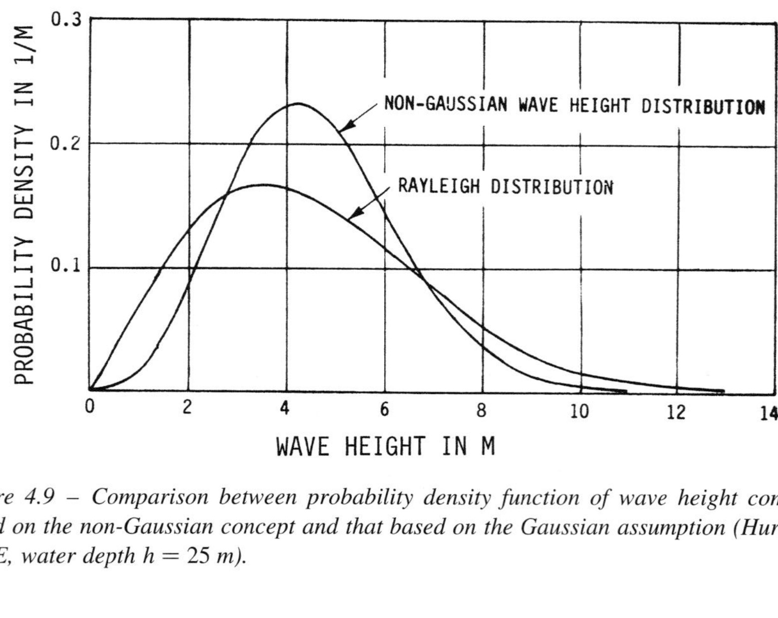

Figure 4.9 – Comparison between probability density function of wave height computed based on the non-Gaussian concept and that based on the Gaussian assumption (Hurricane KATE, water depth h = 25 m).

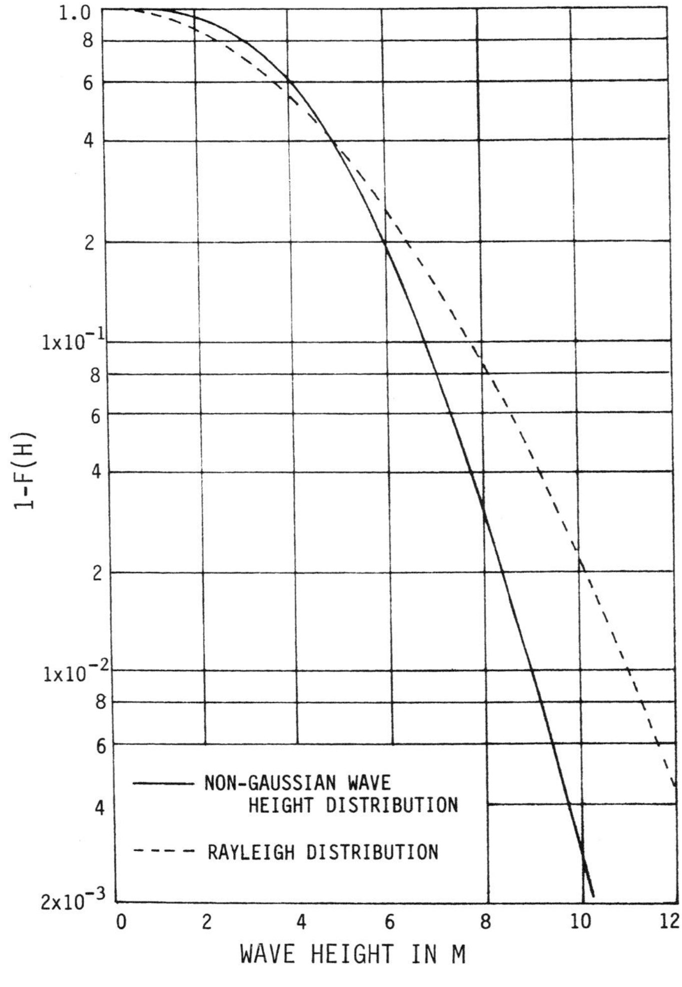

Figure 4.10 – Comparison between probability of exceeding a specified wave height computed based on the non-Gaussian concept and that based on the Gaussian assumption (Hurricane KATE, water depth $h = 25$ m).

Figure 4.11 taken from a paper authored by Forristall (1978) shows a comparison of $1 - F(H)$ between the Rayleigh distribution and observed storm data taken in the Gulf of Mexico fitted by the Weibull distribution. The Rayleigh distribution substantially overestimates the observed large wave heights. It is mentioned in Forristall's paper that the data show a definite excess of high crest points and a lack of low trough points. This indicates that the measured waves must be non-Gaussian waves. Further discussion on this subject will be given later in reference to the statistical distribution of wave height in shallow water.

It may be well to note here the significance of water depth on the statistical characteristics of wave height. It is often mentioned that the Rayleigh probability distribution

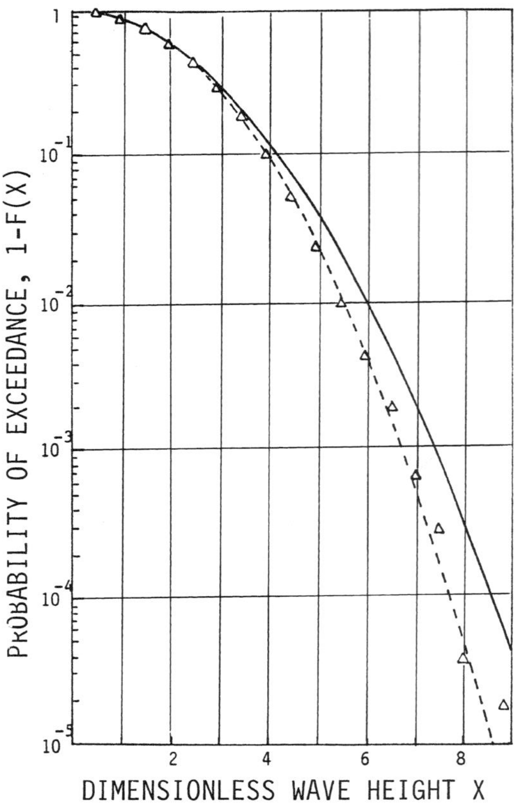

Figure 4.11 – Comparison of probability exceeding a specified wave height computed by Rayleigh distribution (solid line) and Weibull distribution (dotted line) (Forristall, 1978).

overestimates the magnitude of large wave heights in the analysis of wave data obtained in severe seas, or it is often claimed that wave peaks are larger than troughs in severe seas. However, care must be given as to whether or not this can be attributed solely to sea severity. More attention should be given to water depth which may be the prime factor contributing to the question at issue.

It should further be noted that wave data recorded by a wave probe attached to fixed (non-floating) offshore structures in finite water depth usually show wave crests that are larger than wave troughs because the reference zero-line shifts due to storm surge and/or tide. Recorded data so obtained provide valuable information for more rational estimation of the wave-induced loading on fixed offshore structures, but the data will not provide accurate wave information unless a correction for the zero-line shift is made.

5

Hurricane landing and nearshore sea severity

Severe offshore sea conditions generated by a hurricane cause serious problems for ships and offshore structures operating in the advancing path of the hurricane; however, even more chaotic status is observed at the time of hurricane landing which gives rise to a devastating impact to the habitation and natural environment. For example, Leadon, Nguyen, and Clark (1998) give a detailed analysis of beach erosion as well as erosion of the dune system caused by hurricane OPAL.

Hurricane landing most often takes place in the western Pacific during the typhoon season (usually June through October) and the western Caribbean and Gulf of Mexico during the hurricane season (usually June through November). As an example of hurricane landing in the Gulf of Mexico, Figure 5.1 shows the tracks of landing on the Florida coast compiled from available references and overlaid onto a map of the state (McClellan, 1999). In the figure, 9-878 and 8-64, for example, designate September 1878 and August 1964, respectively. Tropical cyclones which landed on the Florida coast and more than 50 hurricanes which passed nearby but did not land on the Florida coast are not included in the figure.

Also included in the figure are two contour lines of water depth for 10 fathoms (18 m) and 100 fathoms (180 m) along the Florida coast. The latter line indicates the approximate distance of the deep water area from the shoreline. The distance from the shore to deep water is unique at each location along the Florida coast. The results of computations of nearshore sea severities at several locations along the Florida coast will be shown later. The results indicate that the severity depends not only on the strength of the hurricane, but is equally dependent on the nearshore bottom configuration where the landing takes place. We will first address the estimation of nearshore sea severity when the hurricane is approaching the shoreline.

5.1 Estimation of nearshore sea severity

Wave characteristics in the nearshore area are complicated in that the effect of extraneous elements such as tide and currents, etc. has to be considered in addition to the reduction in bottom depth. Numerous studies on nearshore waves and sea conditions have been carried out

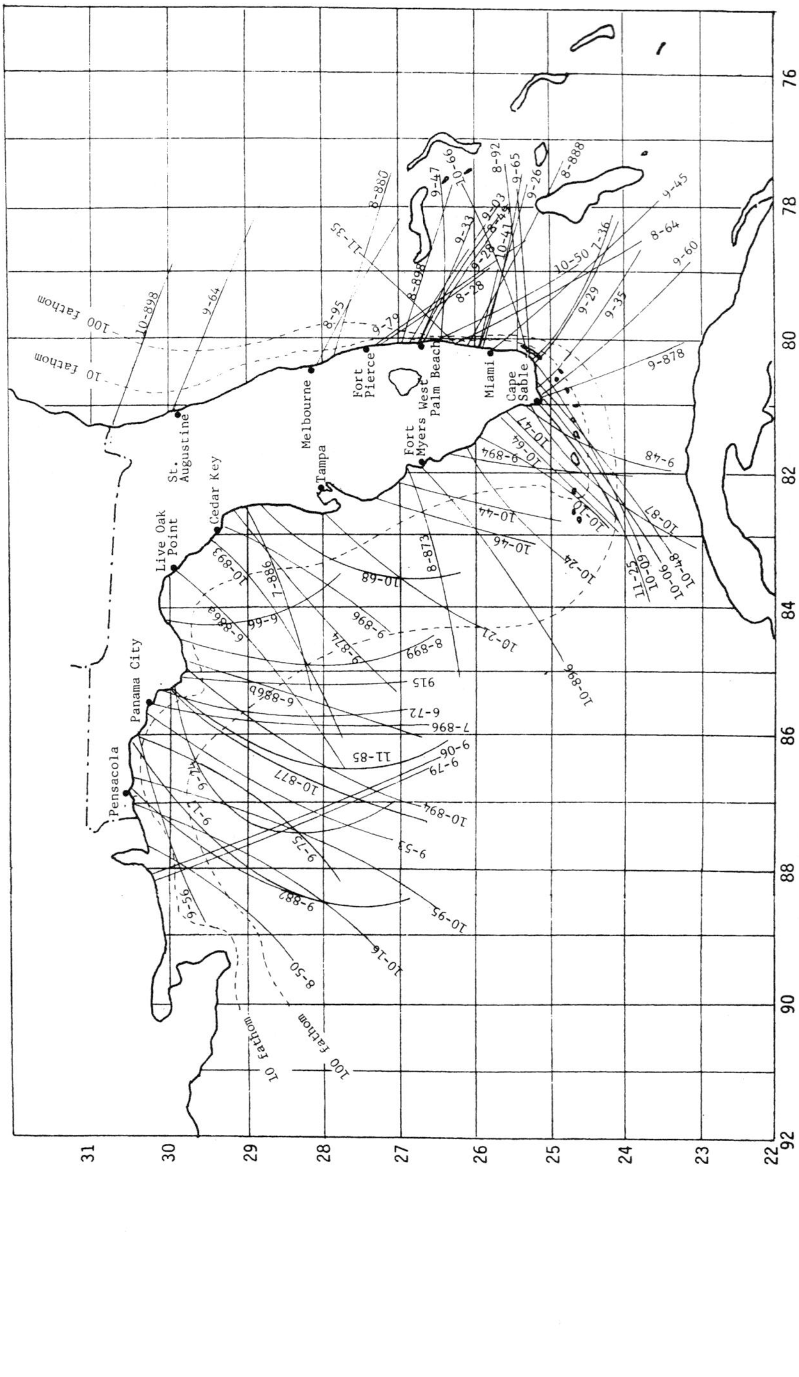

Figure 5.1 – Hurricanes known to have landed on the Florida coast (McClellan, 1999).

by Mizuguchi (1982), Thornton and Guza (1983), Freilich and Guza (1984), Cox, Kobayashi, and Wurjanto (1992), Hamm, Madsen, and Peregrine (1993), among others. A state-of-the-art in the area of nearshore waves is given by Battjes (1994). As mentioned earlier, however, no wave data recorded during hurricane landing are currently available. Hence, one way to estimate the nearshore wave condition at the time of hurricane landing is to assess the sea state from the results of analysis carried out on nearshore wave data obtained in severe storms. In this respect, data obtained during the ARSLOE Project by the Coastal Engineering Research Center are extremely valuable in that wave data were measured simultaneously at 20 min intervals at various water depths ranging from 10 to 2.3 m during a severe storm.

Let us first examine shallow water wave characteristics from the view point of a non-Gaussian random process based on analysis of measured data in the ARSLOE project. Figure 5.2 shows a comparison of the histogram of wave height constructed from data at a water depth of 2.3 m with the probability density function of wave height based on the non-Gaussian random process concept presented in the Appendix. The wave data were taken for 13 min during which 177 waves were recorded. The parameters of the non-Gaussian distribution are estimated from knowledge of the variance (area under the spectrum), the water depth ($h = 2.3$ m) and the calm water depth at the recording site ($D_0 = 1.34$ m). The agreement between the histogram and the probability density function observed in Figure 5.2 is satisfactory.

Significant wave height evaluated from the measured data is 1.78 m, while that estimated from the non-Gaussian distribution is 1.86 m. The extreme wave height recorded in 177 waves is 2.65 m, while the estimated probable extreme height is 2.75 m. Thus, the estimated heights are both on the order of 4–5% greater than the measured heights in this

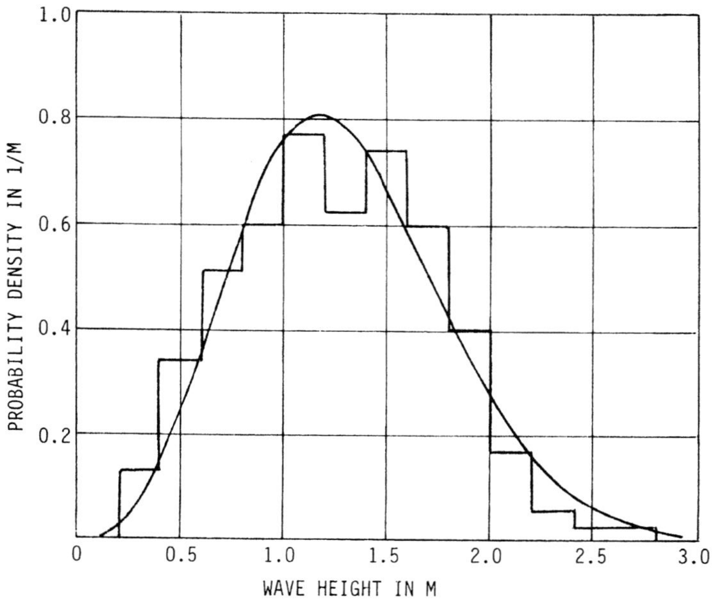

Figure 5.2 – Comparison of histogram of wave height constructed from data obtained in shallow water (2.3 m depth) with the computed non-Gaussian probability density function (Data from ARSLOE Project).

example. If estimation is made under the assumption of Gaussian waves, the estimated significant wave height is 1.86 m which, in this case, agrees with that predicted assuming non-Gaussian waves, but the probable extreme wave height in 177 waves is 3.00 m which is much greater than the observed height.

Figure 5.3 shows the probability that wave height exceeds a specified value H; namely, $1 - F(H)$, where $F(H)$ is the cumulative distribution function of wave height computed assuming non-Gaussian waves (solid line) as well as Gaussian waves (dotted line). The latter is the Rayleigh probability distribution. Also included in the figure is the probability obtained from the measured data shown in Figure 5.2. As seen, the estimated probability computed assuming the non-Gaussian wave height distribution agrees reasonably well for 80% of the data, but overestimates the large wave heights. On the other hand, the Rayleigh probability distribution which is applicable for Gaussian waves deviates from the measured data over the entire range of wave height. The wave height at the crossing point of the

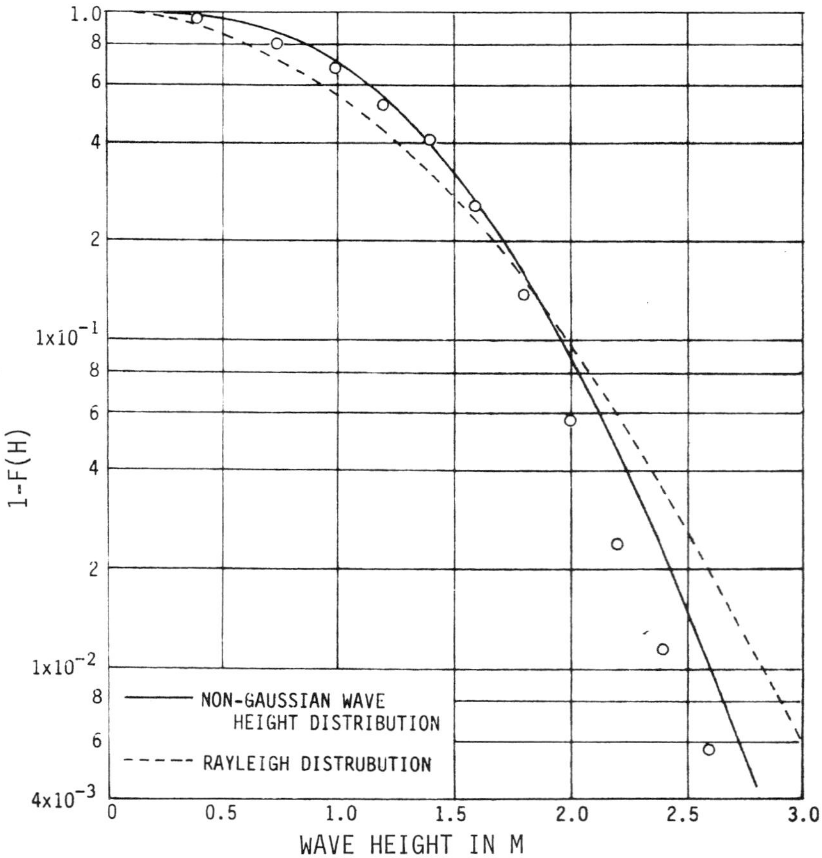

Figure 5.3 – Comparison between probability of exceeding a specified wave height obtained from measured data (open circle) and the probability computed based on the non-Gaussian concept (solid line) and Gaussian assumption (dotted line).

Rayleigh and non-Gaussian distributions is coincidentally equal to the measured significant wave height 1.78 m in this example.

Figure 5.4 shows a comparison between the computed probability density function of peaks and troughs and the associated histograms constructed from the measured data. Since the magnitude of troughs is small, the histogram consists of only six groups and thereby the χ^2-test cannot be performed. The overall agreement, however, between the histograms and the non-Gaussian probability density functions of amplitude is satisfactory. The probability density function demonstrates a significant difference between peaks and troughs in shallow water (non-Gaussian) waves; therefore, the wave characteristics (including extreme values) of

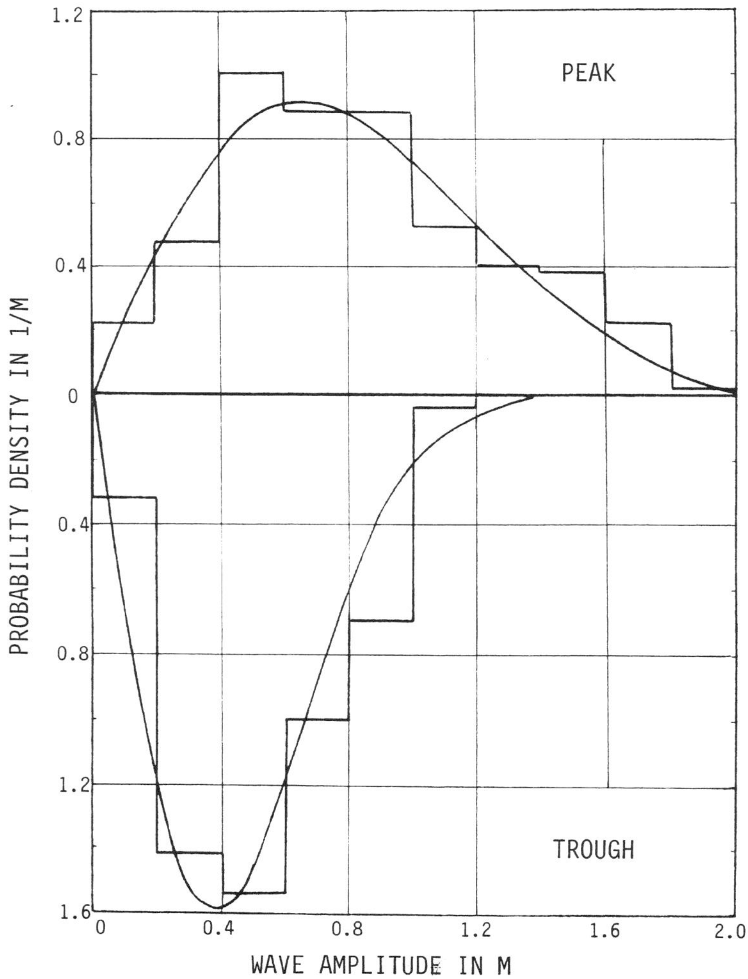

Figure 5.4 – Comparison between peak and trough amplitude histograms constructed from measured data and the non-Gaussian probability density functions (Data from ARSLOE Project).

hurricane-generated seas in the surf zone should be estimated by applying the non-Gaussian probability distribution function to peak and troughs separately.

It may be of interest to see how intense the nearshore waves of hurricane-generated seas is in comparison with those in ordinary severe storm seas. Figure 5.5 shows a comparison between the probability density function of wave height computed in hurricane KATE and that obtained during the ARSLOE Project (Gage #615) at the same water depth of 2.3 m. The computed $\sqrt{m_0}$ (where m_0 is the area under the spectrum) of hurricane KATE is 0.583 m, while that measured by Gage #615 is 0.466 m. This implies that the sea severity of hurricane KATE is approximately 1.25 times of that measured in ordinary severe storm seas. However, analytically there is no linear relationship between the two probability density functions shown in Figure 5.5 because they are not Rayleigh probability distributions. For example, extreme wave height expected in 30 min observation in the sea generated by hurricane KATE is not 1.25 times of that in ordinary severe storm seas.

Figure 5.6 shows the non-Gaussian probability density function of wave height computed in hurricane KATE for various nearshore water depths. The probability density functions are evaluated from the wave spectrum at each water depth estimated from the broken wave spectrum at the breaking point 8.3 m.

The probability density functions of peaks and troughs computed in hurricane KATE for various water depths; 8.3, 6, 4, 2, and 1 m are shown in Figure 5.7. The probability density function at 8.3 m depth is that for waves after energy is lost due to breaking. The probability density function of wave troughs at 1 m water depth is truncated at 1 m because of the depth limitation. As seen in the figure, the shapes of the amplitude distribution of peaks and troughs are significantly different, and the rate of change of the density function rapidly increases as

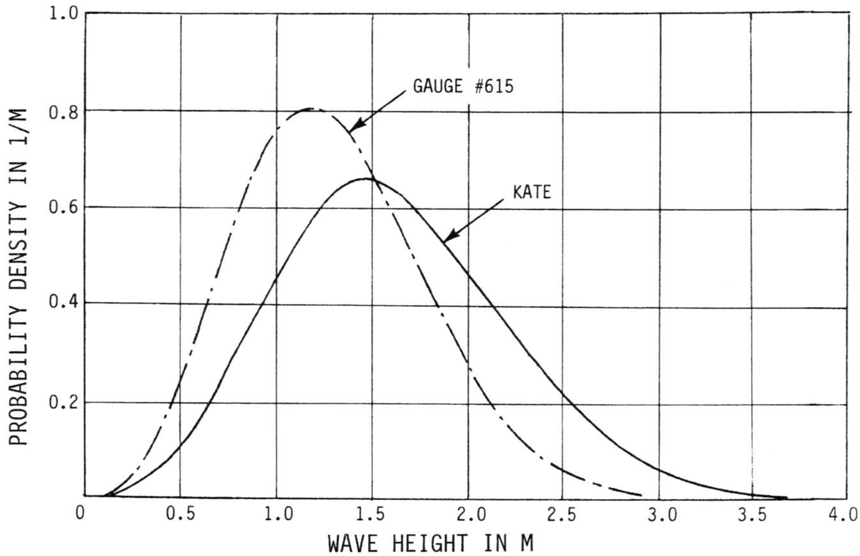

Figure 5.5 – Comparison between probability density function of wave height computed in hurricane KATE and that computed in a severe storm in ARSLOE Project for the same water depth 2.3 m.

the waves approach the shoreline. In particular, the rate of change of trough distributions is pronounced since wave troughs are much more susceptible to bottom effect.

The probability density functions of wave amplitudes and heights presented so far are those computed for the water level being equal to the calm water depth at each location during the landing. This situation may be observed when the hurricane landing takes place at the time of low tide so that the magnitude of low tide and that of storm surge cancel each other. In reality, the water depth may most likely be greater than the calm water depth at the time of hurricane landing.

The effect of increase in sea level on the nearshore sea severity is significant at the time of hurricane landing. There are several mathematical models for estimating storm surge, but this subject is beyond the scope of this text. Furthermore, estimation of the magnitude of tide at the time of hurricane landing is unpredictable. Hence, in the following, the effect of water level on nearshore sea severity at the time of hurricane landing is evaluated for a specified change in sea level representing the sum of storm surge and tidal change without identifying the contribution of each part.

In evaluating the sea severity (significant wave height) in shallow water, it is important to consider whether the sea level is rising or receding since the non-Gaussian properties of waves are different in these two stages. It is noted that an onshore flow is observed in the former stage, while an offshore flow occurs in the latter stage in which the direction of flow is counter to the approaching hurricane.

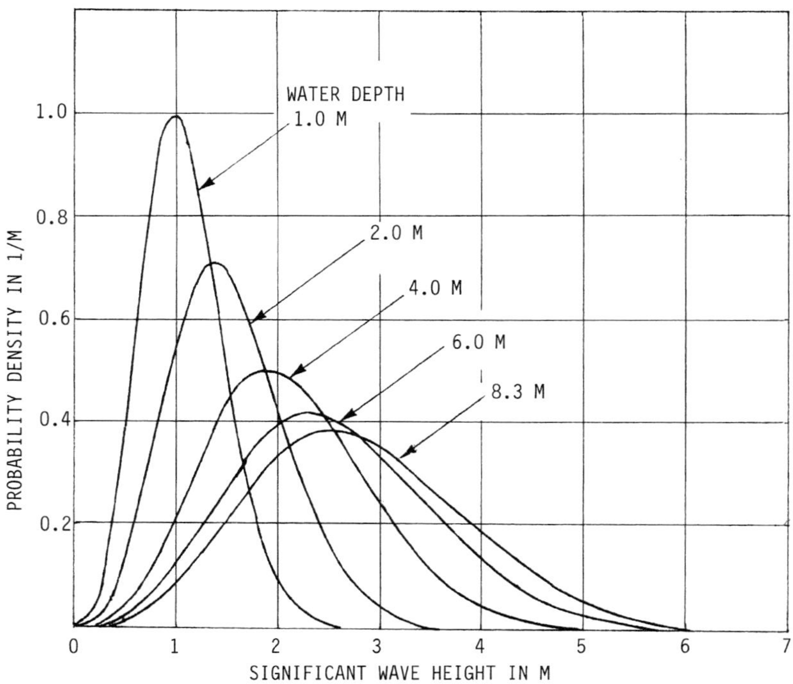

Figure 5.6 – Non-Gaussian probability density function of wave heights computed in hurricane KATE for various nearshore water depths.

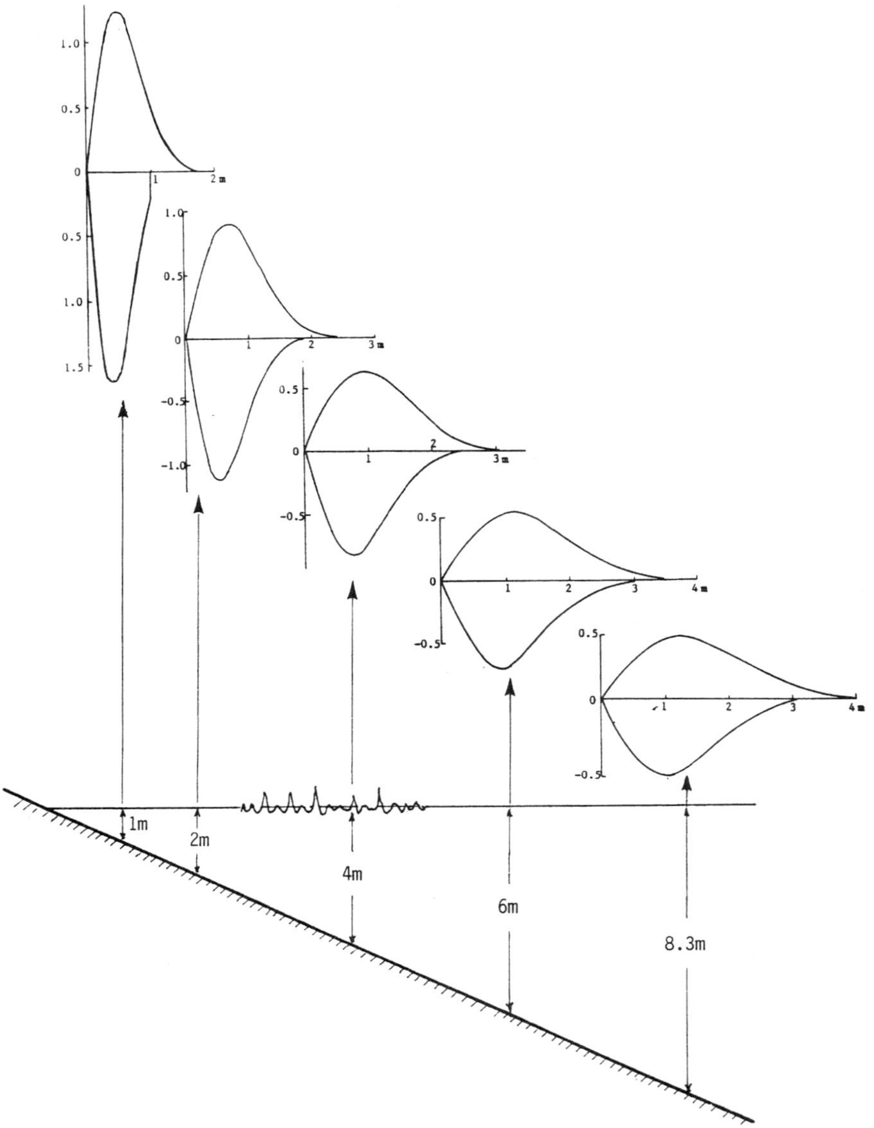

Figure 5.7 – Probability density function of peak and trough amplitudes computed in hurricane KATE for various nearshore water depths.

Figure 5.8(a) shows significant wave height as a function of water depth when hurricane KATE approaches the shoreline assuming that the sea level is equal to the calm water level, while Figure 5.8(b) shows the significant wave height when the sea level has increased by 2 m above the calm water level at the time of landing. In the latter case, the beach and dune beyond the normal shoreline are underwater. The sea severity for a specified on-land water depth is the same as the sea severity for the same water depth of the sea in the absence of

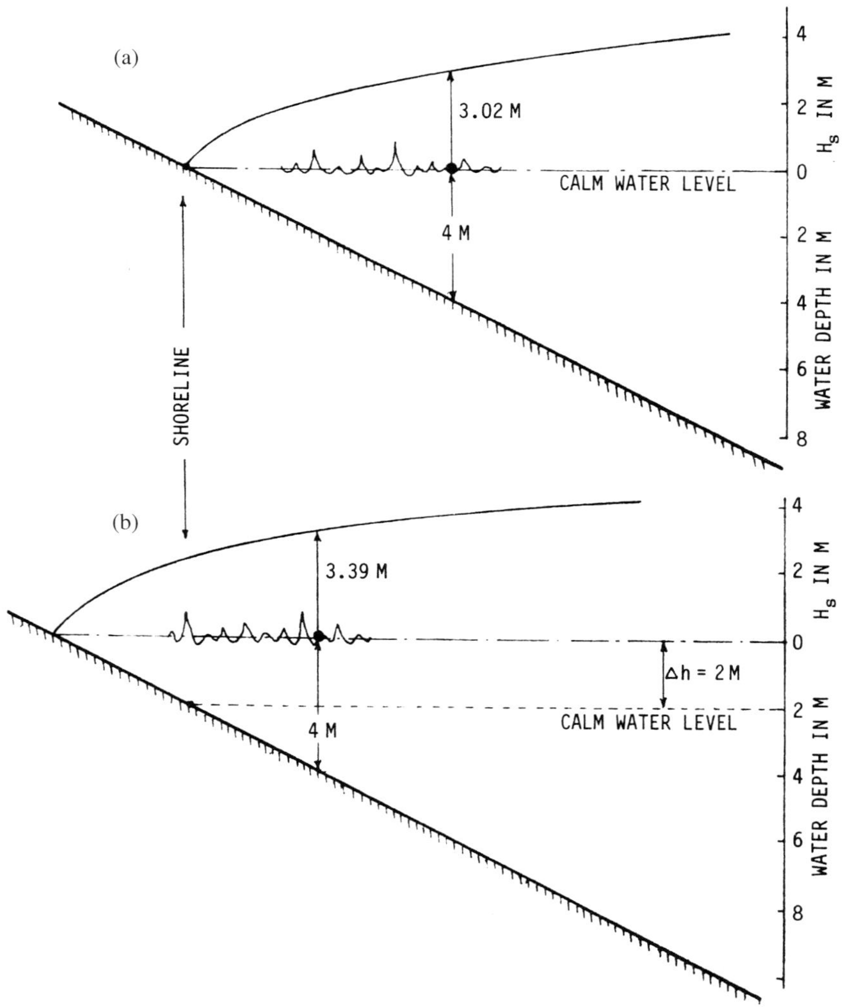

Figure 5.8 – Effect of rising sea level on nearshore significant wave height computed in hurricane KATE: (a) calm water sea level; (b) rising sea level by 2 m.

storm surge/tide. Let us examine the difference in significant wave height at the water depth of 4 m in the figure. As seen in Figure 5.8(a), significant wave height is 3.02 m when the sea level is equal to the calm water level, while it increases to 3.39 m (about 12.3% increase) when the sea level is raised 2 m above the calm water level.

When the sea level is raised by 1 m above the calm water level, the significant wave height is 3.14 m; about a 4% increase in severity for the same water depth of 4 m. Thus, it is clear that the sea severity is increased during the rising stage of water level associated with storm surge/tide. This feature of increase in sea severity associated with uprising water level, however, is observed only in the nearshore area; the increase in severity is almost nil at the breaking point.

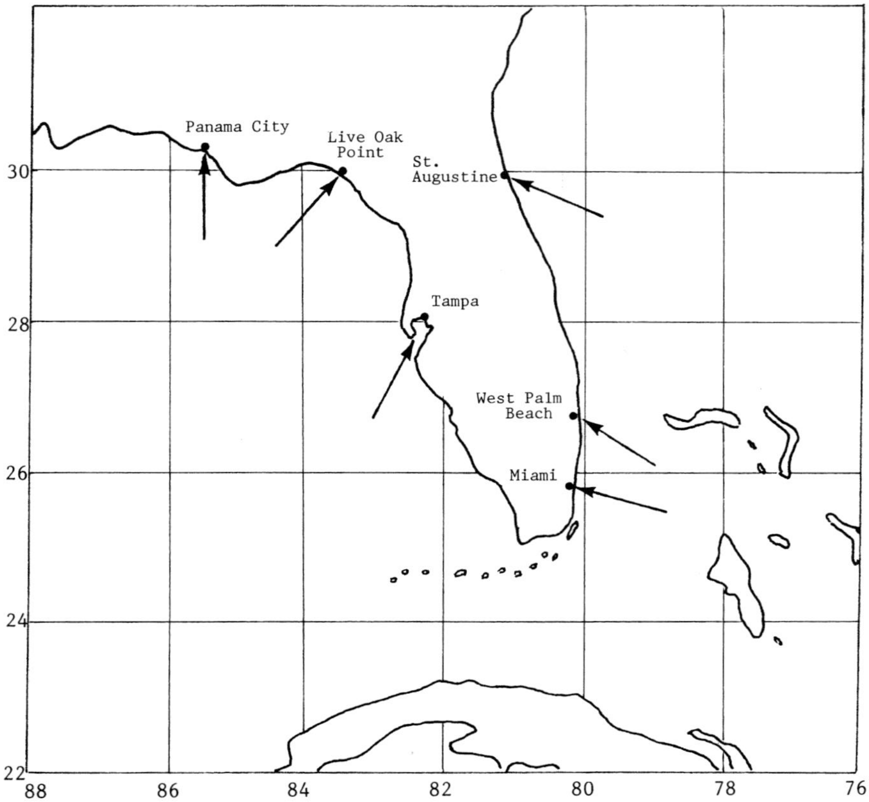

Figure 5.9 – Six locations along the Florida coast chosen for evaluating nearshore severity at the time of hurricane landing.

It should be noted that evaluation of sea severity for a relatively large increase in water level, Δh, above a low calm water level D_0 is not possible. This is because the chart for evaluating the parameter a of non-Gaussian waves given in the Appendix has limits in the value of h/D_0 and σ/h, where h is the water depth, D_0 the calm water depth and σ the square root of the area under the wave spectrum. The computation at $h = 4$ m with $\Delta h = 2$ m shown in Figure 5.8(b) is made by estimating h/D through extending the limit value of 1.7 to 2.0, but the computation yields a result which appears to be reasonable in this case.

Next, let us examine the effect of bottom profile (slope) on nearshore sea severity when a hurricane is close to the shoreline. A study on this subject is carried out by McClellan (1999) at various locations along the Florida coast. In the following, his computation results are shown with minor modifications incorporated by employing an improved method developed later for estimating the nearshore sea severity.

As demonstrated in Figure 5.1, a considerable number of hurricanes have landed at various locations along the Florida coast. Here, six locations along the coast as shown in Figure 5.9 are chosen for the discussion of the effect of bottom profile. Also included in the figure is the direction most likely to be taken in landing at or near that location.

The nearshore (distance less than 6 km) bottom profiles along the specified direction at these locations are shown in Figure 5.10. It is noted that Figure 5.1 gives an impression that the bottom profiles near Miami and West Palm Beach are very steep. However, if consideration is given to the bottom profile in the nearshore area (say, within a distance of 1 km), it is seen from Figure 5.10 that the bottom profile at Panama City is more sharp than at Miami and West Palm Beach.

Based on the information given in Figure 5.10, the significant wave height is computed for various water depths assuming hurricanes having the same strength as KATE land at the six locations independently, and the results are shown in Figure 5.11 as a function of distance from the shoreline at each site. In these computations, the average water depth is assumed to be that of the calm water level.

From comparison of Figures 5.10 and 5.11, it is found that the order of sequence of nearshore sea severity (significant wave height) in Figure 5.11 follows the order of nearshore bottom steepness observed in Figure 5.10. Thus, it may be stated that when a hurricane is close to the shoreline, the nearshore sea severity depends largely on the bottom profile in the neighborhood of 500–800 m off the shoreline where the water depth is less than 10 m along the Florida coast. This conclusion is derived based on the results of computations made on hurricane KATE, but the same trend is also obtained in the results of computations carried out on hurricanes ELOISE and GLORIA (McClellan, 1999).

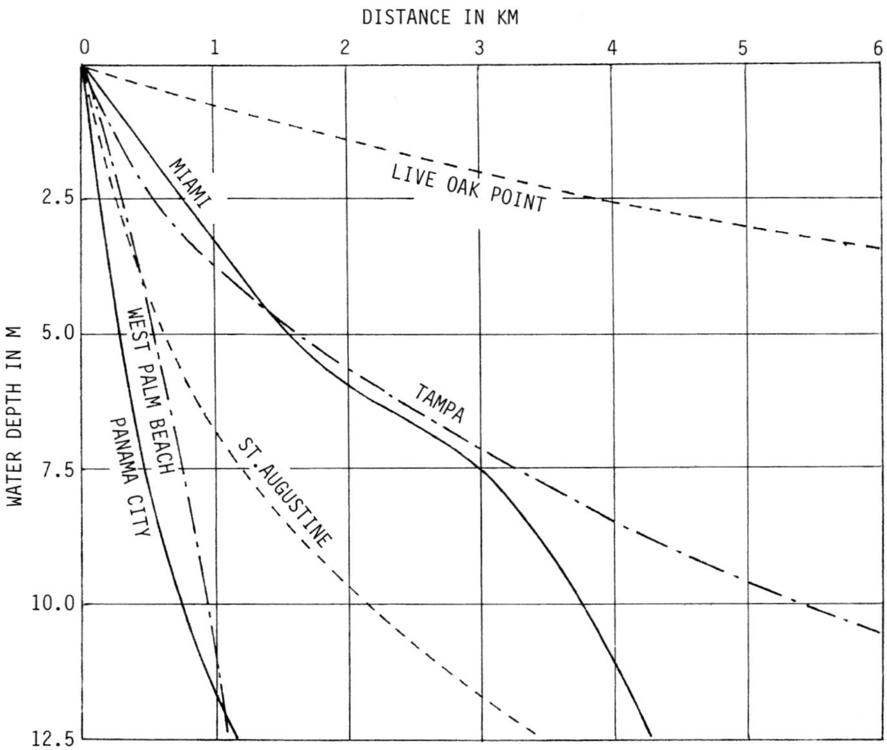

Figure 5.10 – Nearshore bottom profile at six locations along the Florida coast (McClellan, 1999).

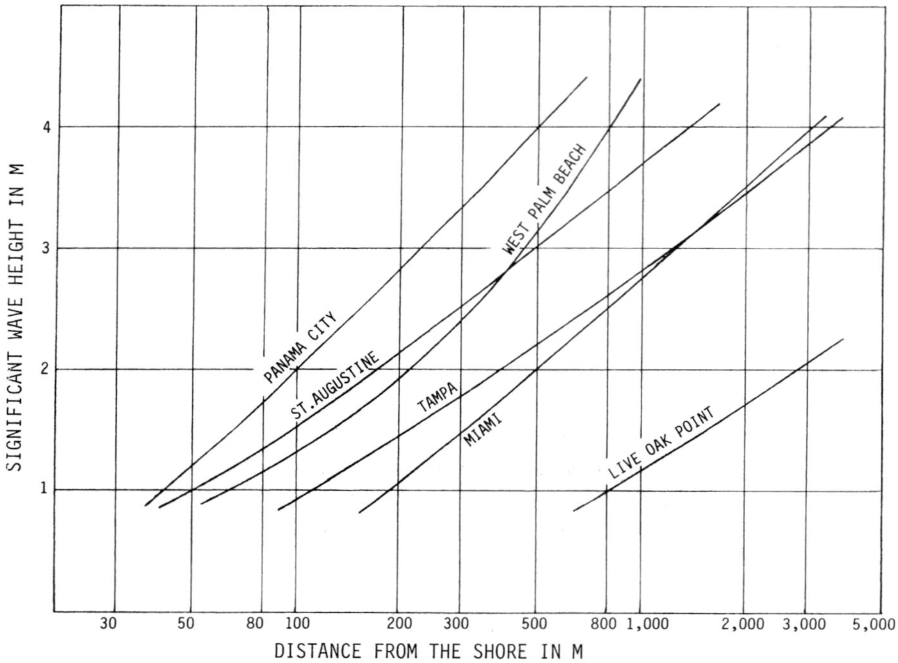

Figure 5.11 – Significant wave height as a function of distance from the shoreline at various locations along the Florida coast assuming hurricanes having the same strength as KATE land at each location (Ochi, Malakar, and McClellan, 2000).

5.2 Hurricane landing sea severity indicator

In Section 5.1, severity of the nearshore sea condition when a hurricane is approaching the shoreline is discussed. General conclusions obtained on the nearshore sea severity may also be applicable at the time of hurricane landing. However, the sea severity at the time of hurricane landing is theoretically zero at the shoreline. It is highly desirable, therefore, to formulate an indicator which reflects a relative severity of the nearshore sea condition at the time of hurricane landing.

One way to represent the sea severity at the time of hurricane landing is to evaluate the rate of reduction in significant height as a function of distance from the shoreline. As an

Water depth, h(m)	8.0	4.0	1.9	0.5
Significant wave height, H_s(m)	4.0	3.0	2.0	1.0
Distance, L(km)	0.50	0.22	0.10	0.04
Sea severity-distance ratio, H_s/L	8.0	13.3	20.0	25.0

Table 5.1 *Sea severity–distance ratio computed at various water depths when hurricane KATE is approaching Panama City*

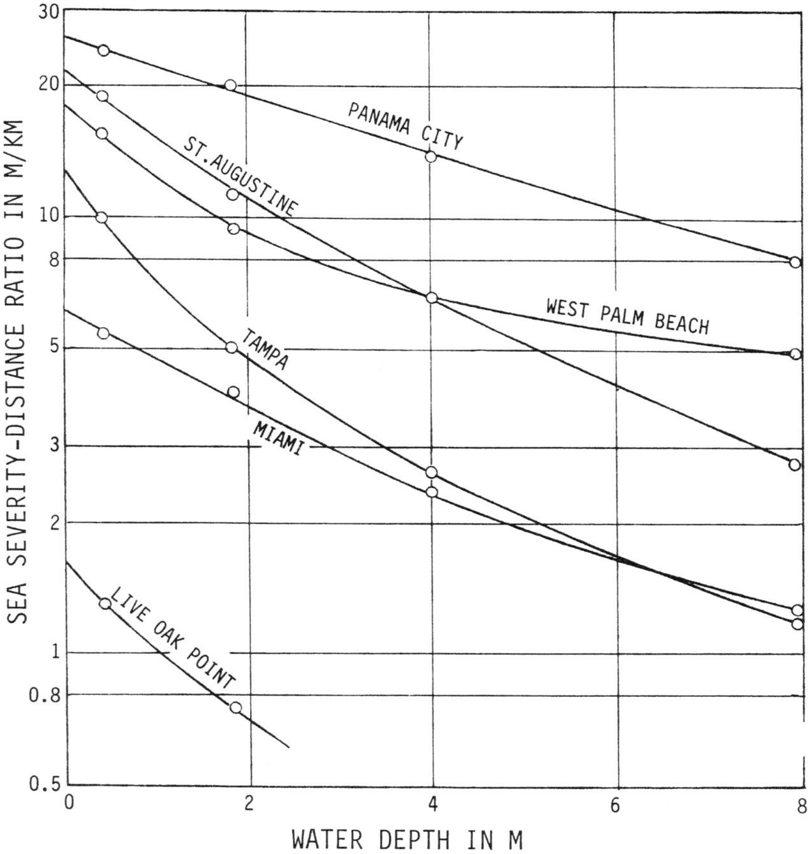

Figure 5.12 – Sea severity–distance ratio computed at six locations assuming hurricanes having the same strength as KATE land at each location (McClellan, 1999).

example, let us consider the case when hurricane KATE is approaching Panama City. The computed significant wave height, distance from the shoreline and the "sea severity–distance ratio", defined as the ratio of significant wave height and the distance from the shoreline, are given in Table 5.1 as a function of nearshore water depth.

As seen in Table 5.1, the sea severity–distance ratio increases as the hurricane approaches the shoreline. The functional relationship obtained for Panama City is shown in Figure 5.12 along with similar relationship lines obtained for an additional five locations along the Florida coast. It is possible to evaluate the value of the ratio at zero water depth (shoreline) by extending each relationship line, and this dimensionless value at the shoreline may be called the "landing sea severity indicator" which represents the relative severity of the sea condition at hurricane landing.

It is clear from Figures 5.10 and 5.12 that the magnitude of the "landing sea severity" indicator depends to a great extent on the nearshore bottom configuration. In general, the indicator is large for a location where the bottom configuration has a large slope in the nearshore area, i.e. the sequence of the order of landing sea severity indicators shown in

Figure 5.12 agrees with the sequence of the order of steepness of nearshore bottom slope observed in Figure 5.10. It may be said from Figure 5.12 that if hurricane KATE had come to Tampa following the approach route shown in Figure 5.9, the nearshore sea severity at the time of landing would be half of that experienced at Panama City.

5.3 Onshore sea severity

Because of storm surge and tide, the beach and part of the dune system will be underwater at the time of hurricane landing, and hence we have to consider the sea severity thereon which may be called the onshore sea severity. Although we cannot control the nearshore sea severity, it is possible to control the onshore sea severity to some extent by arranging a proper slope of the dune system.

For computing the extent of the portion of land underwater due to storm surge/tide, let the slope of the nearshore bottom profile be θ_0, continuing for a distance l on the beach to a dune system (represented here by a straight line) at an additional angle $\Delta\theta$ as illustrated in Figure 5.13. Point A in the figure is the shoreline for the calm water level, Δh is the combined magnitude of water level associated with storm surge/tide. The magnitude of Δh could be negative when the hurricane lands at low tide, but we consider Δh here to be positive, since the storm surge at the time of hurricane landing is likely to be much greater than the magnitude of the tide at the landing site. In this case, the dune system will be underwater only when the magnitude of the rising water level Δh is greater than $l \sin\theta$. Then, the underwater portion of the dune system, BC in Figure 5.13, can be evaluated by

$$\mathrm{BC} = \frac{\Delta h - l \sin\theta_0}{\sin(\theta_0 + \Delta\theta)} \qquad \text{where } \Delta h \geq l \sin\theta_0 \tag{5.3-1}$$

The total underwater length on land, denoted by L_*, becomes

$$L_* = \mathrm{BC} + l = \frac{\Delta h}{\sin(\theta_0 + \Delta\theta)} + l\left\{1 - \frac{\sin\theta_0}{\sin(\theta_0 + \Delta\theta)}\right\} \tag{5.3-2}$$

As an example, we evaluate the underwater portion of the land when hurricane KATE lands at Panama City assuming that the nearshore bottom slope θ_0 is approximately 3° ($\sin\theta_0 = 0.05$). Computations are made for three increased water levels (Δh =1, 2 and 3 m), six beach depths from the calm water shoreline ($l = 0$ through 50 m) and three slopes of the dune system ($\Delta\theta = 1, 2$ and 3°). In the case when $\Delta\theta = 0$, the dune system is merely an extension of the beach. The results of the computations are shown in Figure 5.14 in which the distance of the

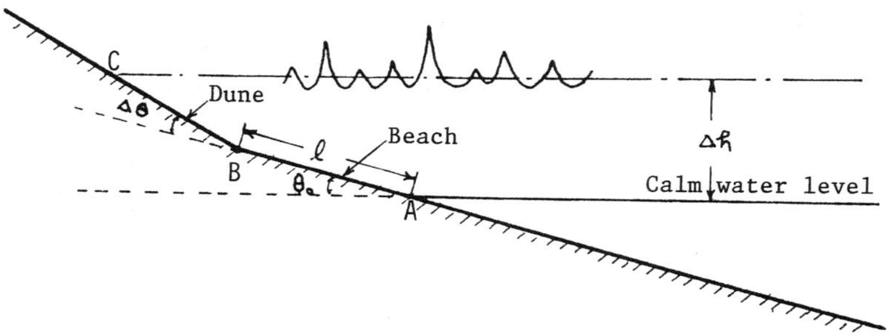

Figure 5.13 – Diagram for computation of onshore sea severity.

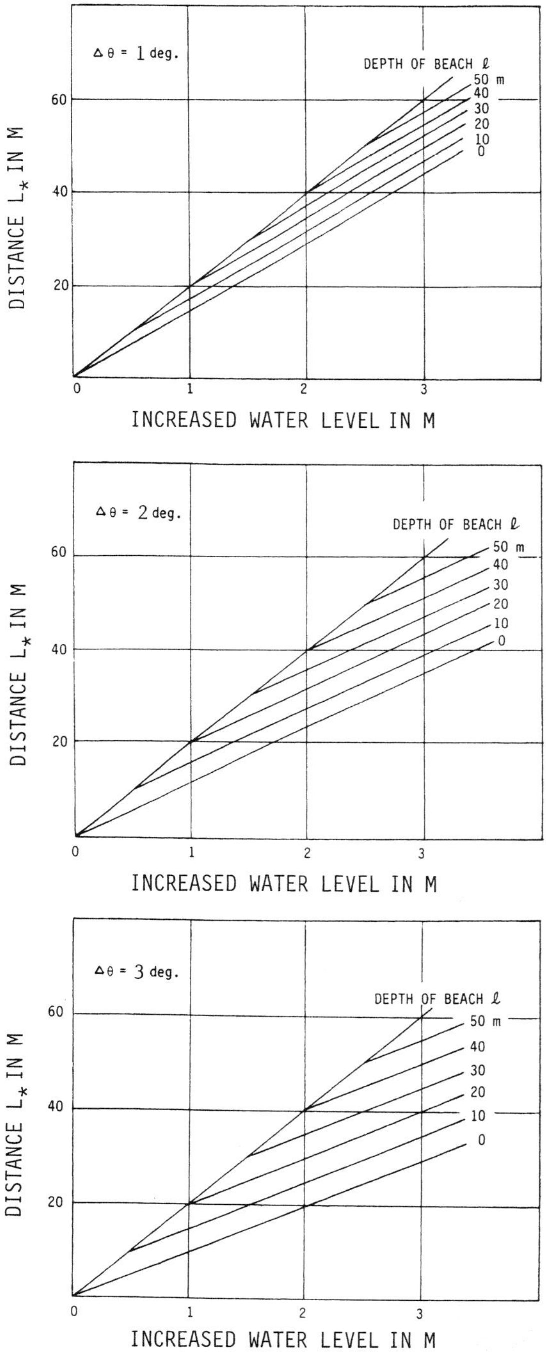

Figure 5.14 – Distance of underwater portion of land, L_*, as a function of increased water level Δh for dune system slope $\Delta\theta = 1, 2$ and $3°$.

underwater portion of the land, L_*, is given as a function of increased water level Δh for a specified depth of the beach l and slope of the dune system $\Delta\theta$.

It is understood from Equation (5.3-2) that for a specified increase in water level Δh there exists a maximum underwater length L_* when $\Delta\theta = 0$ (no dune) which yields the maximum underwater length $L_* = \Delta h / \sin\theta_0$. This is given by the diagonal straight lines in Figure 5.14 in which $\sin\theta_0 = 0.05$.

Let us evaluate a possible reduction in underwater distance from the maximum length L_* for the nearshore bottom slope $\theta_0 = 3.0°$ ($\sin\theta_0 = 0.05$). As an example, let the specified increase in water level Δh be 2 m. In this case, the maximum L_* is 40 m from Figure 5.14 and a reduction in the magnitude of L_* cannot be expected if the beach depth l is greater than 40 m. A reduction of L_* can be achieved by choosing a proper combination of the beach depth and the dune slope. In other words, if the beach length $l = 30$ m and the dune slope $\Delta\theta = 2°$, the underwater length L_* becomes 36 m, while if $l = 20$ m and $\Delta\theta = 3°$, the underwater length L_* is reduced to 30 m. These computations are made assuming the beach configuration remains constant.

As another example, let the depth of the beach l be specified as $l = 30$ m. Figure 5.15 shows the effect of the slope of the dune system $\Delta\theta$ on the underwater portion L_* as a function

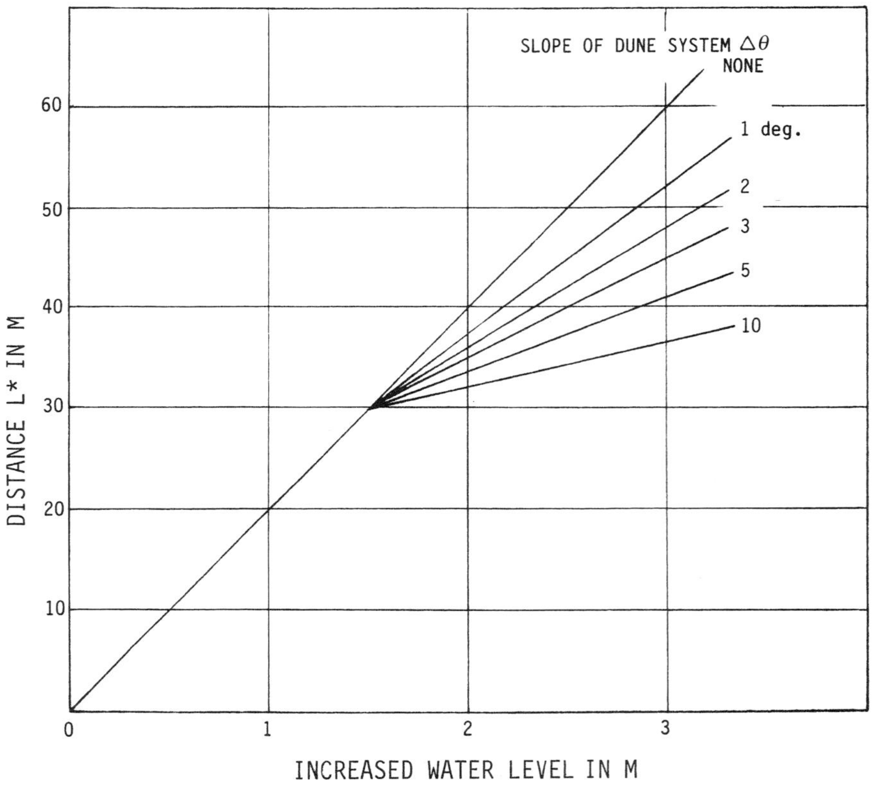

Figure 5.15 – Effect of slope of dune system, $\Delta\theta$, on distance of underwater portion of land (Extent of beach $l = 30$ m).

of increase in water level Δh. Let the magnitude of rising water level Δh be 3 m. In this case, the dune system will not be underwater unless the magnitude of the rising water is greater than 1.5 m because the depth of the beach l is 30 m. If a 3° angle is given to the dune system, the underwater portion of the dune system can be reduced by 25% (from 60 to 45 m) in comparison with no slope, while an angle of 10° given to the dune system will result in a reduction of 38%.

Thus, for a given nearshore bottom slope θ_0, a reduction of underwater length is feasible for a specified magnitude of increase in water level associated with storm surge and or tide, Δh, by choosing an appropriate depth of beach, l, and angle of dune system $\Delta\theta$.

It may be of interest to evaluate the magnitude of onshore sea severity; significant wave height, for example. The magnitude of significant wave height in onshore sea severity is the same as that in the nearshore for the calm water level depth. The significant wave height in shallow water when hurricane KATE is approaching the shore is computed for non-Gaussian waves and the results are shown in Figure 5.16. With the aid of Figures 5.15 and 5.16, an example of the onshore sea severity (significant wave height) when hurricane KATE lands at Panama City is shown in Figure 5.17. It is assumed in this example that the beach depth from the calm water shoreline is 30 m and that the magnitude of rising water due to storm surge/tide at the time of landing is 3.0 m. As seen, an increase in the angle of the dune system is beneficial in that it reduces not only the underwater area but also the onshore sea condition to a great extent.

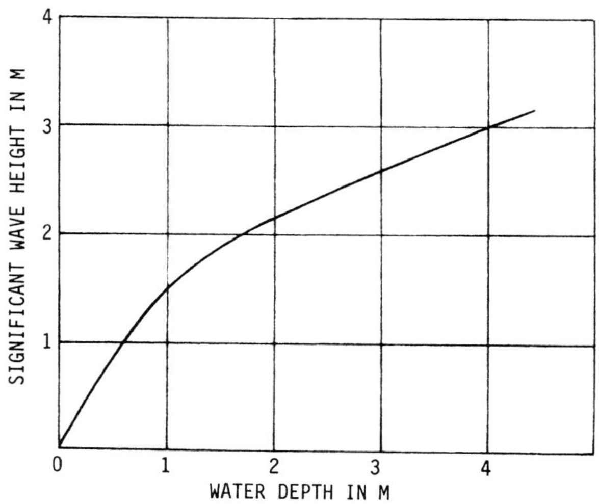

Figure 5.16 – Reduction in significant wave height when hurricane KATE approaches shoreline.

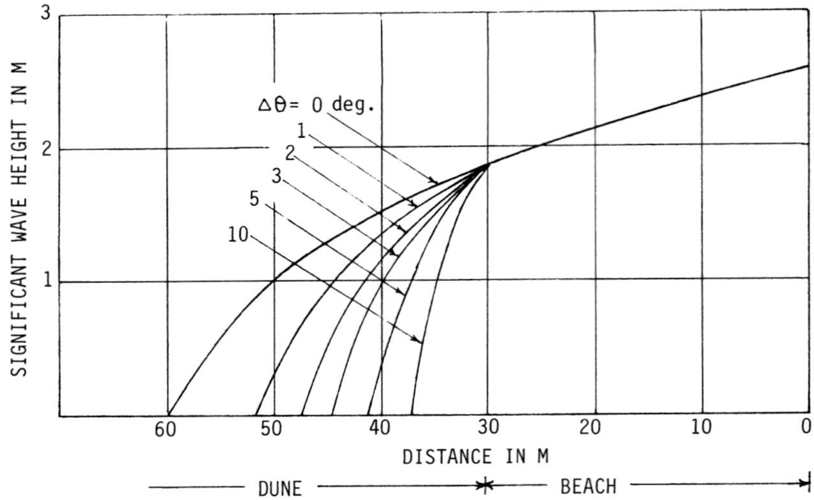

Figure 5.17 – Expected onshore sea severity when hurricane KATE lands at Panama City (Beach depth $l = 30$ m, uprising water level $\Delta h = 3$ m).

Appendix

Review of non-Gaussian random waves

A.1 Probability distribution of wave profile

The time history of the profile (displacement from the mean value) of waves in finite water depth indicates the randomly changing profile of the positive and negative parts to be dissimilar, in general (see wave record in Figure A.1), which is in contrast to that observed for waves in deep water. For waves in deep water, it can be analytically proved by the central limit theorem that the probability distribution of the wave profile obeys the normal (Gaussian) probability distribution. Consequently, random waves in deep water are called Gaussian waves, while waves in finite water depth are considered to be non-Gaussian random waves.

Probability distributions applicable for presenting non-Gaussian random waves have been derived by three different approaches: (a) application of orthogonal polynomials to the probability density function (Longuet-Higgins, 1963); (b) application of Stokes wave theory (Tayfun, 1980; Huang, *et al*, 1983); and (c) application of the Kac–Siegert solution to a non-linear system (Langley, 1987; Ochi and Ahn, 1994a,b).

The non-Gaussian probability density function derived by applying the concept of orthogonal polynomials is known as the Gram–Charlier series probability density function. The probability density function is given in series form and thereby usually has a negative value at some part of the distribution caused by the finite number of waves used in computing the density function.

The probability density function derived based on the application of Stokes waves imposes a preliminary form on the wave profile for derivation of its probability distribution such that the individual waves are expressed as a Stokes expansion to the second or third order components. Furthermore, the Stokes expansion is valid for waves having $kh > 1$, where k is the wave number and h water depth. Hence, there is some reservation in applying Stokes' theory for random waves in finite water depth, particularly shallow water waves, unless the expansion includes higher-order terms.

On the other hand, Kac and Siegert (1947) analytically show that a non-linear system whose response is presented in the form of a two-term Voltera's stochastic series expansion

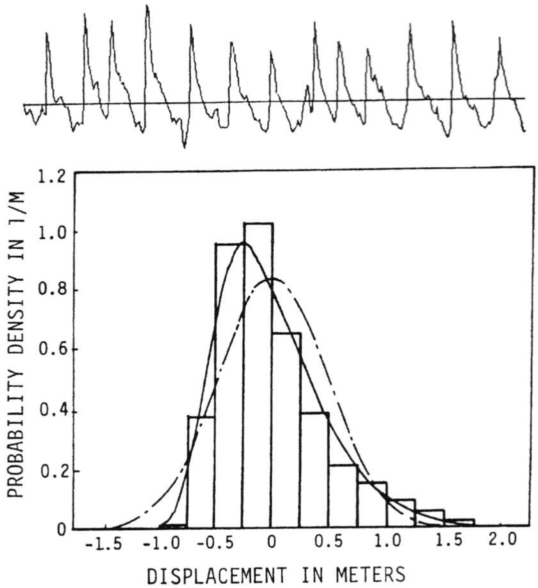

Figure A.1 – Comparison between wave profile (displacement) histogram constructed from data obtained in shallow water and non-Gaussian probability density function (solid line) and Gaussian probability density function (broken line).

can be presented in terms of the standardized normal random variable as follows

$$y(t) = \sum_{j=1}^{n} (\beta_j z_j + \lambda_j {z_j}^2) \tag{A.1-1}$$

where $y(t)$ is the output of the non-linear system and z_j the standardized normal variate.

The parameters β_j and λ_j are evaluated by finding the eigen function and eigen values of the integral equation consisting of a second-order frequency response function and the output spectral density function. This approach to obtain the probability density function of a non-linear system is analytically precise and application of this concept to non-linear random waves is made by Langley (1987). That is, in order to obtain β_j and λ_j in Equation (A.1-1) from the wave record, the parameters are determined based on the concept of wave–wave interaction by applying second-order wave potential theory. The functional relationship between the interaction coefficients and the parameters are presented in a concise matrix form.

In order to derive the probability density function of waves in finite water depth with strong non-linearity, Ochi and Ahn (1994a) evaluate the non-linear wave–wave interaction by decomposing the wave spectrum into linear and non-linear components with the aid of bispectral analysis and the results are incorporated in Langley's matrix presentation. Comparison between the probability density function so derived and the histogram constructed from non-Gaussian wave data shows good agreement. Thus, the Kac–Siegert approach yields a valid probability density function for non-Gaussian random waves. However, the probability density function is not given in closed form and thereby the probabilistic prediction of other wave properties such as wave amplitude and height cannot be derived from it.

In order to derive the probability density function in closed form Ochi and Ahn (1994a) present Equation (A.1-1) as a function of a single random variable U as

$$Y = U + aU^2 = \left\{\sqrt{a}\left(U + \frac{1}{2a}\right)\right\}^2 - \frac{1}{4a}, \tag{A.1-2}$$

where U is a normal variate with mean μ_* and variance ${\sigma_*}^2$, and a is a constant. These quantities are determined by equating the cumulant generating functions of Equations (A.1-1) and (A.1-2). Here, the cumulant generating function is defined as the logarithm of the characteristic function and it can be expanded into a series in the form of

$$\psi(t) = \sum_{s=1}^{\infty} k_s \frac{(it)^s}{s!} = k_1(it) + \frac{1}{2!}k_2(it)^2 + \frac{1}{3!}k_3(it)^3 + \cdots \tag{A.1-3}$$

where k_s refers to the cumulants.

From comparison of each term of the cumulant generating functions of Equations (A.1-1) and (A.1-2), we have

$$\begin{aligned} &a{\sigma_*}^2 + a{\mu_*}^2 + \mu_* = 0 \\ &\sum_j {\beta_j}^2 + 2\sum_j {\lambda_j}^2 = {\sigma_*}^2 - 2a^2{\sigma_*}^4 \\ &3\sum_j {\beta_j}^2\lambda_j + 4\sum_j {\lambda_j}^3 = a{\sigma_*}^4(3 - 8a^2{\sigma_*}^2) \end{aligned} \tag{A.1-4}$$

Here, the left-side of the second and third equations of Equation (A.1-4) are expressed by cumulants as

$$\begin{aligned} &\sum_j {\beta_j}^2 + 2\sum_j {\lambda_j}^2 = k_2 = E[y^2] - (E[y])^2 \\ &3\sum_j {\beta_j}^2\lambda_j + 4\sum_j {\lambda_j}^3 = \tfrac{1}{2}k_3 = \tfrac{1}{2}\{E[y^3] - 3E[y^2]E[y] + (E[y])^2\} \end{aligned} \tag{A.1-5}$$

where $E[y^j]$ are the jth moment of the data. Thus, by letting the mean value of the wave data be the zero-line, k_2 and k_3 are equal to the sample moments $E[y^2]$ and $E[y^3]$, respectively, and thereby the unknown parameters, a, μ_*, and σ_* are determined from Equation (A.1-4).

Upon evaluating the three parameters, the probability density function of Y can be obtained from the density function of U by applying the technique of change of random variables. However, the density function so derived vanishes at $y = -(1/4a)$ due to a singularity involved in the density function. To overcome this difficulty, the functional relationship between U and Y is expressed as

$$U = \frac{1}{\gamma a}(1 - e^{-\gamma a Y}) \tag{A.1-6}$$

where γ is a constant: 1.28 for $y \geq 0$, and 3.00 for $y < 0$. These values are determined from analysis of the functional relationship between non-dimensionalized Y and U, and are valid for non-Gaussian waves with very strong non-linear characteristics.

It should be noted that the sample space of the normal variate U defined in Equation (A.1-6) is $(-\infty, 1/\gamma a)$ instead of $(-\infty, \infty)$. However, the truncation does not affect the probability distribution in practice, since the truncation point $1/\gamma a$ is much larger than $\mu_* + 3\sigma_*$ of the normal distribution with mean μ_* and variance ${\sigma_*}^2$.

By using the functional relationship given in Equation (A.1-6), the probability density function of the displacement Y can be derived from the probability distribution of U which is a normal probability distribution, i.e.

$$f(y) = \frac{1}{\sqrt{2\pi}\sigma_*}\exp\left\{-\frac{1}{2(\gamma a\sigma_*)^2}(1-\gamma a\mu_* - e^{-\gamma ay})^2 - \gamma ay\right\} \quad (A.1\text{-}7)$$

where γ is 1.28 for $y \geq 0$ and 3.00 for $y < 0$.

It is noted that the values of γ are different for positive and negative y-values. This results in a slight difference in the slope of the probability density function at $y = 0$, although the density function is continuous at this point. It is also noted that the density function

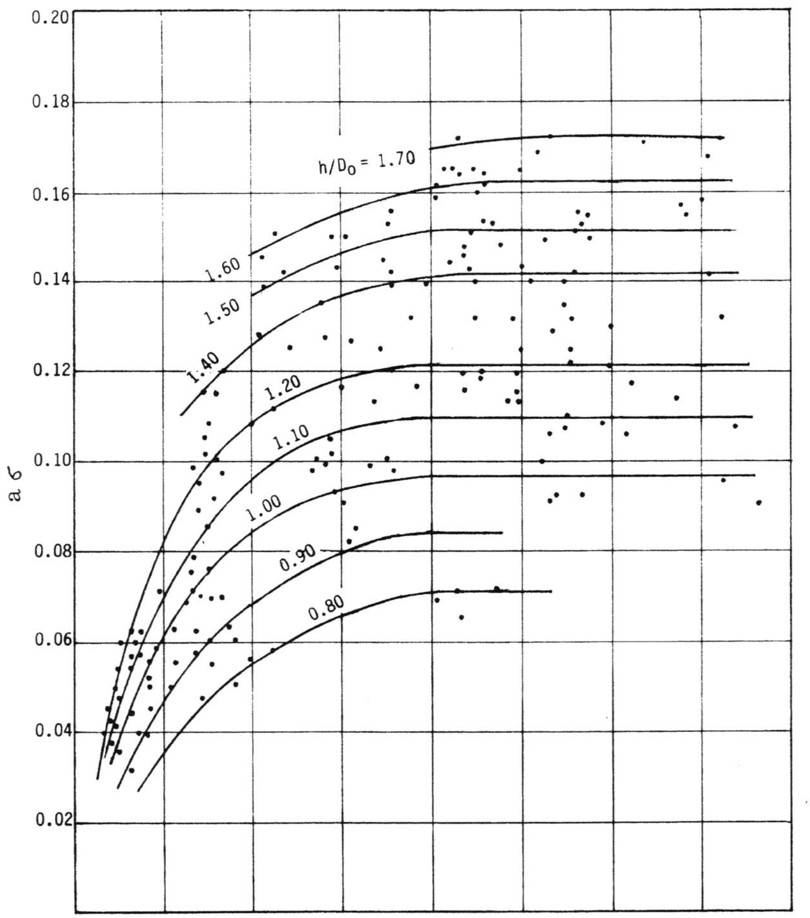

Figure A.2 – Parameter $a\sigma$ plotted against σ/h for various h/D_0 (Water rising situation, Data from ARSLOE Project, Coastal Engineering Research Center).

reduces to a normal distribution if $a = 0$ and that the integrated value of the density function over its sample space is unity.

Comparison between the probability density function given in Equation (A.1-7) and the histograms computed from measured data agree very well for several hundred wave data obtained from deep to shallow water during the ARSLOE Project (Ochi and Ahn, 1994b). An example of the comparison for shallow water data at a water depth of 2.3 m with strong non-linearity is shown in Figure A.1.

As discussed above, the parameters a, μ_*, and σ_* involved in the probability density function of non-Gaussian waves are determined from wave data. It is extremely important in practice that these parameters be evaluated from the wave spectrum or from knowledge of the area under the wave spectrum. For this, Robillard and Ochi (1996) apply the non-Gaussian probability distribution formula given in Equation (A.1-7) to more than 1000 samples of wave data in finite water depth obtained by the Coastal Engineering Research Center during the ARSLOE Project, and analyzed the results such that the parameters are presented as a function of water depth and sea severity. The conclusions of their study are summarized in the following which includes the results of additional analysis more recently carried out.

The parameter a which represents the degree of wave non-linearity is a function of not only the water depth h during the storm, but also the variation of water level from the calm water depth D_0 at the location. Further, it may be necessary to consider the case when either the water level is rising due to flood tide and/or storm surge or the water level is receding due to ebb tide. In the case of the former, the water is moving shoreward, while in the latter case the water is flowing offshore. Hereafter, the parameter a is presented in the dimensionless form $a\sigma$, where σ is the square-root of the area under the wave spectrum.

Figure A.2 shows the relationship between the parameter $a\sigma$ representing the severity of wave non-linearity and the sea severity–water depth ratio σ/h as a function of the

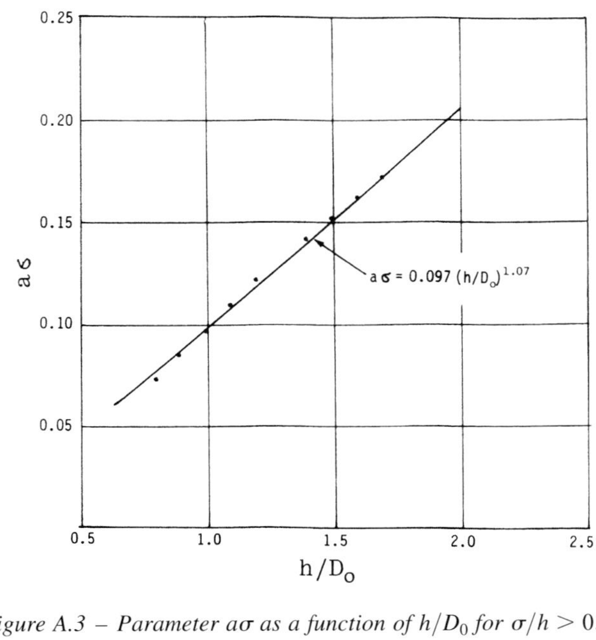

Figure A.3 – Parameter $a\sigma$ as a function of h/D_0 for $\sigma/h > 0.12$.

dimensionless ratio h/D_0. The figure pertains to the rising water situation in which the water is moving shoreward. As seen, for a specified water depth status h/D_0, the value of $a\sigma$ increases sharply with σ/h between 0.06 and 0.12 beyond which it remains constant. For $\sigma/h > 0.12$, the parameter $a\sigma$ can be presented as a function of h/D_0 as

$$a\sigma = 0.097(h/D_0)^{1.07} \tag{A.1-8}$$

This relationship is shown in Figure A.3.

A relationship similar to that given in Figure A.2, but under the receding water situation is also evaluated and a comparison of the results with the rising water situation (Figure A.2) is shown in Figure A.4. As seen, there is no appreciable difference between them at any water depth. This implies that the effect of water flow whether it is moving shoreward or moving offshore is not significant on the statistical characteristics of non-Gaussian waves. Instead, the effect of variation of water depth from the calm water level appears to be pronounced.

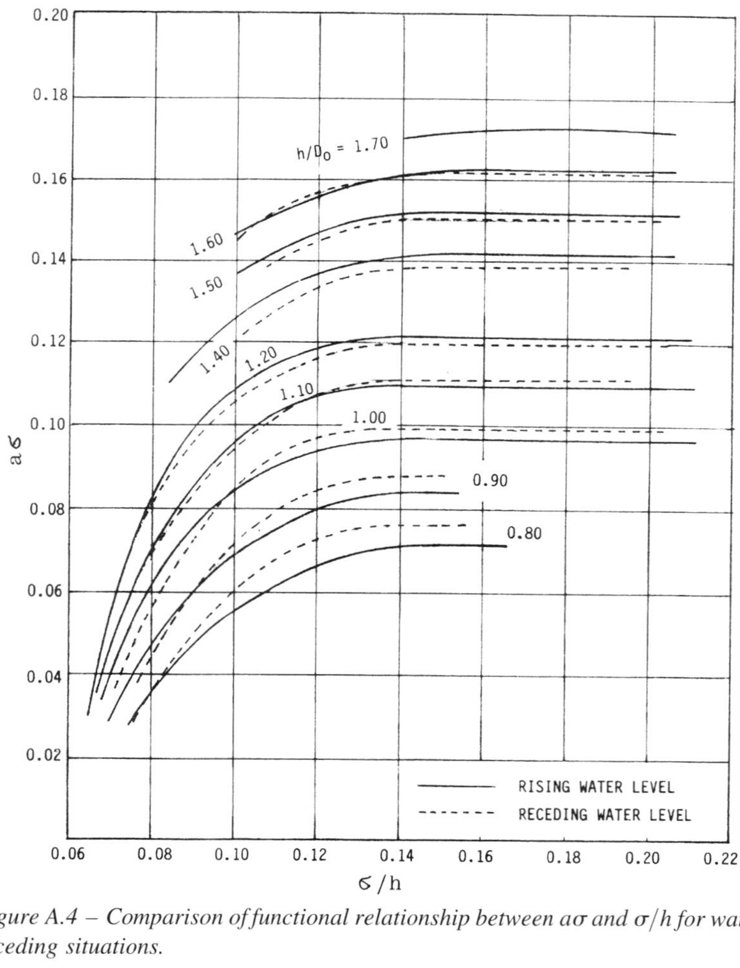

Figure A.4 – Comparison of functional relationship between $a\sigma$ and σ/h for water rising and receding situations.

The parameter μ_* is presented in non-dimensional form μ_*/σ and the functional relationship between μ_*/σ and $a\sigma$ is shown in Figure A.5. As seen, μ_*/σ can be approximately expressed as a function of $a\sigma$ as follows

$$\mu_*/\sigma = -1.55(a\sigma)^{1.20} \tag{A.1-9}$$

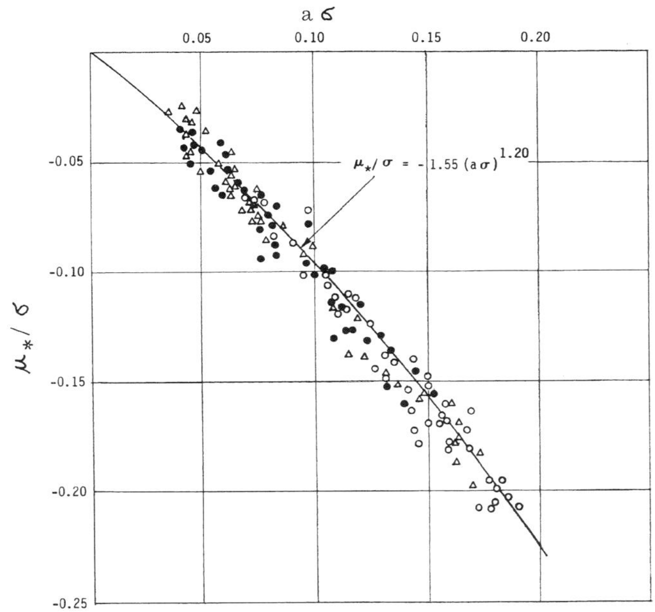

Figure A.5 – Functional relationship between μ_*/σ and $a\sigma$ (Robillard and Ochi, 1996).

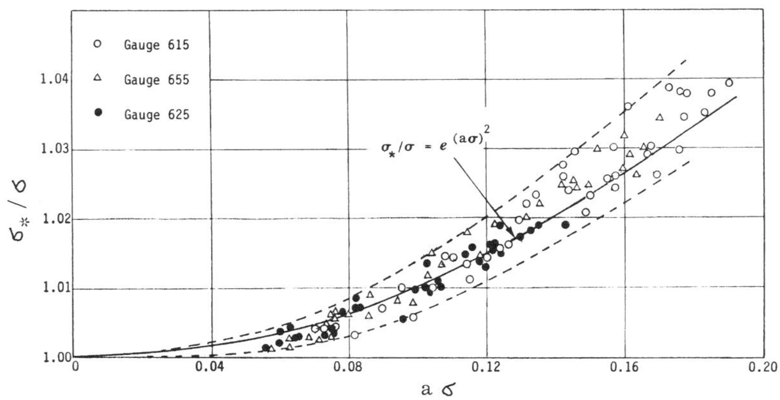

Figure A.6 – Functional relationship between σ_*/σ and $a\sigma$ (Robillard and Ochi, 1996).

Figure A.6 shows the functional relationship between σ_*/σ and $a\sigma$. The vertical scale range of the figure is very small; therefore, the difference between the upper and lower bounds of the scattered data is very small—within 1%. The functional relationship can be presented as follows

$$\sigma_*/\sigma = \exp\{a\sigma)^2\} \qquad \text{(A.1-10)}$$

Thus, in summary, the three parameters a, μ_*, and σ_* requisite for evaluating the statistical properties of non-Gaussian waves can be determined from Equations (A.1-4) and (A.1-5) if wave records are available. If the wave spectrum or the area under the spectral density function is known, then the three parameters can be approximately evaluated from Equations (A.1-8) through (A.1-10) with the aid of Figures A.2 and A.3.

A.2 Probability distribution of wave amplitude and height

For the probability distribution of non-Gaussian waves, distributions of peaks and troughs should be developed separately, and it is highly desirable that the probability distributions are analytically germane to that of the wave profile (displacement). It is noted that the Rayleigh probability distribution is analytically derived for amplitudes of Gaussian waves based on the normal probability distribution for displacements under the narrow-band assumption. A procedure similar to that for Gaussian waves should also be considered for non-Gaussian waves rather than derive a distribution by fitting data with an existing probability distribution. In this respect, only a few probability distributions have been developed applicable for non-Gaussian waves. The probability density function of the amplitude of non-Gaussian waves presented here is analytically developed based on the narrow-band assumption from the density function of wave displacement given in Equation (A.1-7) (Ochi, 1998). The derivation is briefly outlined as follows.

For the probability distribution of peaks, there exists a peak envelope process because of the narrow-band assumption (see Figure A.7). The variance of this process, denoted by ${\sigma_1}^2$, can be evaluated from Equation (A.1-7) as

$${\sigma_1}^2 = 2\int_0^\infty y^2 f(y)\mathrm{d}y = \frac{1}{{\lambda_1}^2}\left\{(\lambda_1\sigma_*)^2 + 2\sqrt{2/\pi}\,(\lambda_1\sigma_*)^3 + 3(\lambda_1\sigma_*)^4\right\} \qquad \text{(A.2-1)}$$

where $\lambda_1 = \gamma a$, and $\gamma = 1.28$ for positive y.

Next, let us define the random variable $V = U - \mu_*$. Here, the random process $V(t)$ is a narrow-band Gaussian process with zero mean and variance ${\sigma_1}^2$. This being the case, $V(t)$ can be decomposed into statistically independent sine and cosine components, denoted by V_c and V_s, respectively, and we derive the joint probability density function $f(V_c, V_s)$. Further, by using the relationships given in Equation (A.1-6) and $V = U - \mu_*$, the joint probability density function $f(V_c, V_s)$ is converted to the joint probability density function $f(Y_s, Y_c)$ by applying the change of random variable technique. Then, we may write the random process $Y(t)$ as

$$Y_c = \xi\cos\tau, \qquad Y_s = \xi\sin\tau \qquad \text{(A.2-2)}$$

where ξ is the amplitude (peak) and τ the phase. After evaluating the joint probability density function of ξ and τ, the probability density function of the peak ξ can be derived as follows

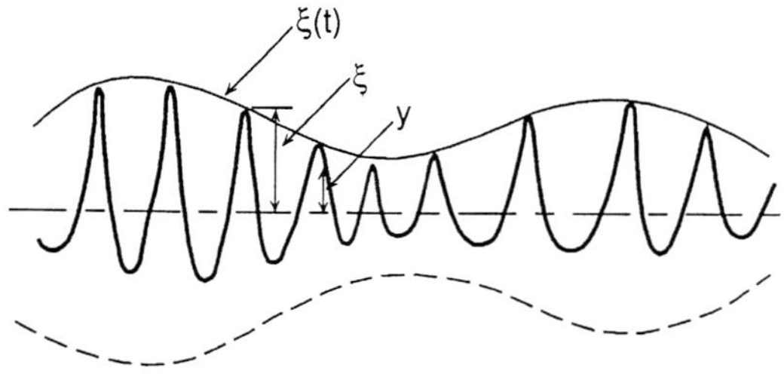

Figure A.7 – Pictorial sketch representing peak envelop process of non-Gaussian waves.

$$f(\xi) = \frac{(1+\lambda_1\mu_*)\xi}{\sigma_1{}^2}\exp\left\{-\left[\frac{(\lambda_1\sigma_1{}^2-\mu_*)^2}{(1+\lambda_1\mu_*)\sigma_1{}^2} + \frac{1+\lambda_1\mu_*}{2\sigma_1{}^2}\xi^2\right]\right\} I_0\left[\sqrt{2}\left(\lambda_1 - \frac{\mu_*}{\sigma_1{}^2}\right)\xi\right] \quad \text{(A.2-3)}$$

In case μ_* and λ_1 are both zero (Gaussian random process), the variance $\sigma_1{}^2$ reduces to σ^2 and the above density function becomes the following Rayleigh distribution.

$$f(\xi) = \frac{\xi}{\sigma^2}\exp\left(-\frac{\xi^2}{2\sigma^2}\right) \quad \text{(A.2-4)}$$

By defining $s_1{}^2$ and c_1 as follows,

$$\frac{\sigma_1{}^2}{1+\lambda_1\mu_*} = s_1{}^2 \quad \text{and} \quad \frac{\sqrt{2}(\lambda_1\sigma_1{}^2-\mu_*)}{1+\lambda_1\mu_*} = c_1 \quad \text{(A.2-5)}$$

Equation (A.2-3) can be presented as

$$f(\xi) = \frac{\xi}{s_1{}^2}\exp\left\{-\frac{1}{2s_1{}^2}(c_1{}^2+\xi^2)\right\} I_0(c_1\xi/s_1{}^2) \quad \text{(A.2-6)}$$

This probability density function has the same form as that developed by Rice (1945) applicable for the sum of two statistically independent processes; one being a narrow-band Gaussian random process with zero mean and variance $s_1{}^2$, the other a sine wave with amplitude c_1; both having the same frequency. This result provides insight as to the structure of the probability distribution of the peak envelope process of non-Gaussian waves.

It is further possible to simplify Equation (A.2-6) for non-Gaussian waves having skewness less than approximately 1.2; the condition observed on almost all non-Gaussian waves. In other words, by approximating the modified Bessel function in Equation (A.2-6) in terms of an exponential function and by normalizing the density function such that the integration in sample space becomes unity, Equation (A.2-6) can be written as

$$f(\xi) = \frac{1}{s_1{}^2}\left(1-\frac{2c_1{}^2}{5s_1{}^2}\right)\xi\exp\left\{-\frac{1}{2s_1{}^2}\left(1-\frac{2c_1{}^2}{5s_1{}^2}\right)\xi^2\right\} \quad \text{(A.2-7)}$$

Equation (A.2-7) is the Rayleigh probability density function with the parameter R_1, i.e.

$$f(\xi) = \frac{2\xi}{R_1}e^{-\xi^2/R_1} \quad \text{(A.2-8)}$$

where $R_1 = 2s_1{}^2/(1-2c_1{}^2/5s_1{}^2)$

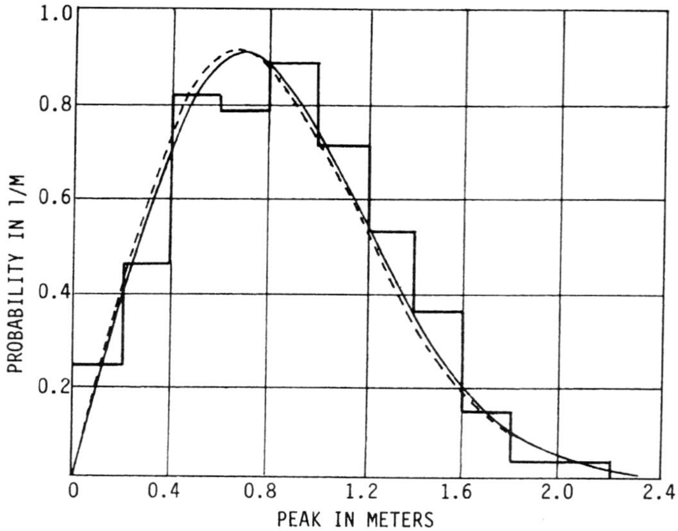

Figure A.8 – Comparison of exact (solid line) and approximate (dashed line) probability density functions applicable for peaks of non-Gaussian waves with the histogram constructed from data (Ochi, 1998).

Thus, it is found that amplitudes of the positive part of non-Gaussian random waves may be approximately distributed following the Rayleigh probability distribution with the parameter R_1. Note that the parameter R_1 is a function of three parameters (a, μ_*, and σ_*) for the non-Gaussian waves.

An example of comparison of the exact (Equation (A.2-6)) and the approximate (Equation (A.2-7)) probability density functions with the histogram constructed from shallow water waves is shown in Figure A.8. As seen, the difference between the exact and approximate density functions is very small and they represent the histogram of peaks very well.

The probability density function applicable to troughs of non-Gaussian random waves, denoted by $f(\eta)$, has essentially the same form as Equation (A.2-7). The parameters s_1 and c_1 in Equation (A.2-7), however, should be replaced by s_2 and c_2, respectively, which are appropriate for troughs. Here,

$$s_2 = \frac{{\sigma_2}^2}{1 + \lambda_2 \mu_*} \quad \text{and} \quad c_2 = \frac{\sqrt{2}(\lambda_2 {\sigma_2}^2 - \mu_*)}{1 + \lambda_2 \mu_*} \tag{A.2-9}$$

where

$${\sigma_2}^2 = 2E[y^2] - {\sigma_1}^2$$

$$\lambda_2 = \gamma a, \qquad \gamma = 3.00.$$

As is the case for the probability distribution of peaks, the density function is approximately expressed by the following Rayleigh probability distribution

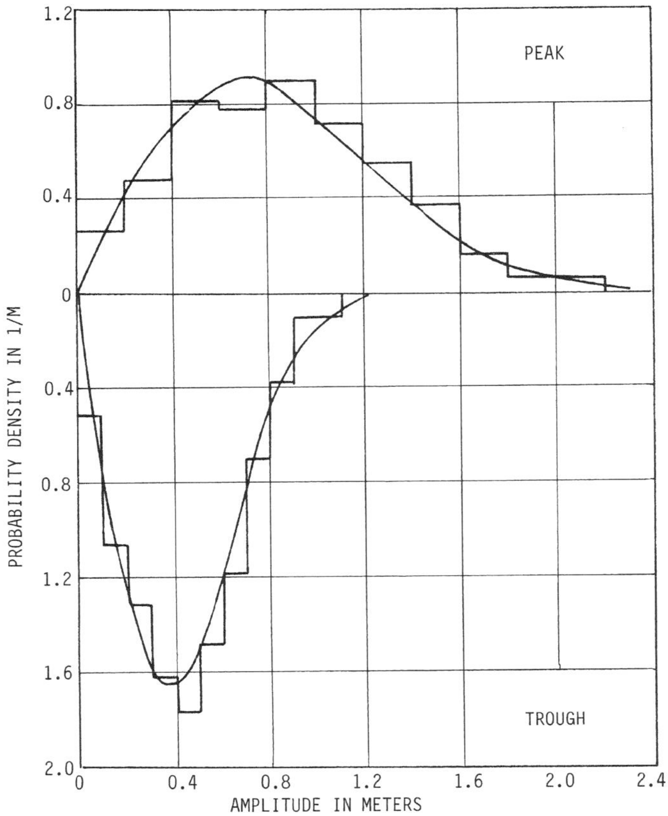

Figure A.9 – Comparison of probability density function of peak and trough amplitudes of non-Gaussian waves with histogram constructed from data (Ochi, 1998).

$$f(\eta) = \frac{2\eta}{R_2} e^{-\eta^2/R_2} \tag{A.2-10}$$

where $R_2 = 2s_2{}^2/(1 - 2c_2{}^2/5s_2{}^2)$

Figure A.9 shows the comparison between the probability density function $f(\xi)$ and $f(\eta)$ along with the histogram of each probability density function constructed from wave data obtained in shallow water. The density function $f(\xi)$ is the same function as shown in Figure A.8. A pronounced difference can be seen in the probability density functions of peaks and troughs, and the computed density function very well represents the histogram on both peaks and troughs.

The probability density function of wave height, denoted by $f(\varsigma)$, can be derived based on the concept that the height is the sum of statistically independent peaks and troughs. It is a convolution integral of the two probability density functions $f(\xi)$ and $f(\eta)$, i.e.

$$f(\varsigma) = \int_0^\infty f_\xi(\xi) f_\eta(\varsigma - \xi) d\xi \tag{A.2-11}$$

The results of computation yields

$$f(\varsigma) = \frac{2\varsigma}{(R_1 + R_2)^2}(R_1\, e^{-\varsigma^2/R_1} + R_2\, e^{-\varsigma^2/R_2})$$

$$+ \frac{2\sqrt{\pi}}{R_1 + R_2} e^{-\varsigma^2/(R_1+R_2)} \sqrt{\frac{R_1 R_2}{R_1 + R_2}} \left(\frac{2}{R_1 + R_2}\varsigma^2 - 1\right)$$

$$\times \left[\Phi\left(\sqrt{\frac{2R_2}{R_1(R_1 + R_2)}}\varsigma\right) - \Phi\left(-\sqrt{\frac{2R_1}{R_2(R_1 + R_2)}}\varsigma\right) \right] \tag{A.2-12}$$

where R_1 and R_2 are the Rayleigh distribution parameters given in Equations (A.2-8) and (A.2-10), respectively, and $\Phi(\,)$ is the cumulative probability distribution of the standardized normal probability function.

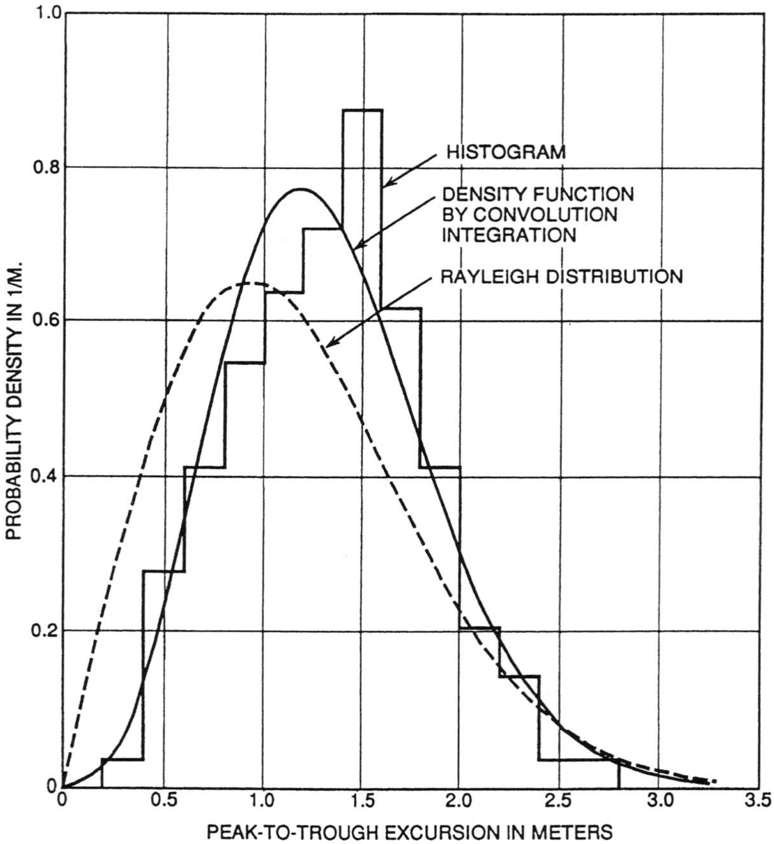

Figure A.10 – Comparison of probability density function of non-Gaussian wave height and histogram constructed from data. Dotted line represents the Rayleigh probability density function based on assumption of Gaussian waves (Ochi, 1998).

Figure A.10 shows the comparison between the probability density function $f(\varsigma)$ and the histogram constructed from the same wave data as used in Figure A.9. Also included in the figure is the probability density function computed based on Gaussian waves; the Rayleigh probability density function. As seen, $f(\varsigma)$ differs significantly from the probability density function computed based on the Gaussian assumption.

The cumulative distribution function of wave height, denoted by $F(\varsigma)$ can be obtained as follows

$$F(\varsigma) = \iint_{\xi+\eta=\varsigma} f(\xi, \eta)\mathrm{d}\xi\mathrm{d}\eta = \int_0^{\varsigma}\left[\int_0^{\varsigma-\xi} f(\eta)\mathrm{d}\eta\right]f(\xi)\mathrm{d}\xi$$
$$= 1 - \frac{1}{R_1+R_2}\left[R_1\,\mathrm{e}^{-\varsigma^2/R_1} + R_2\,\mathrm{e}^{-\varsigma^2/R_2} + 2\varsigma\sqrt{\frac{\pi R_1 R_2}{R_1+R_2}}\,\mathrm{e}^{-\varsigma^2/(R_1+R_2)}\right.$$
$$\left.\times\left\{\Phi\left(\sqrt{\frac{2R_2}{R_1(R_1+R_2)}}\varsigma\right) - \Phi\left(-\sqrt{\frac{2R_1}{R_2(R_1+R_2)}}\varsigma\right)\right\}\right] \tag{A.2-13}$$

References

Antani, J.K. (1981) "Mathematical representation of hurricane associated wave spectra" *University of Florida Report UFL/COEL-81-007.*

Atkinson, G.D. & Holliday, C.R. (1977) "Tropical cyclone minimum sea level pressure maximum sustained wind relationship for western north Pacific" *Monthly Weather Rev.* **105**, 421-427.

Battjes, J.A. (1994) "Shallow water wave modeling" *Proceedings of the International Symposium: Waves Physical and Numerical Modeling*, University of British Columbia, pp. 1-23.

Battjes, J.A. & Beji, S. (1992) "Breaking waves propagating over a shoal" *Proceedings of the 23rd Coastal Engineering Conference*, vol. 1, pp. 42-50.

Battjes, J.A. & Janssen, J.P.F.M. (1978) "Energy loss and set-up due to breaking of random waves" *Proceedings of the 16th Coastal Engineering Conference*, vol. 1, pp. 569-587.

Battjes, J.A., Eldeberky, Y. & Won, Y.S. (1993) "Spectral Boussinesq modeling of breaking waves" *Proceedings of the Ocean Wave Measure Analysis*, pp. 813-820.

Battijes, J.A. & Stive, M.J.F. (1985) "Calibration and verification of a dissipation model for random breaking waves" *J. Geophys. Res.* **90**(C5), 9159-9167.

Bea, R.G. (1974) "Gulf of Mexico hurricane wave height" *Proceedings of the Sixth Offshore Technical Conference OTC 2110.*

Black, J.L. (1979) "Hurricane ELOISE directional wave energy spectra" *Proceedings of the 11th Offshore Technical Conference OTC 3594.*

Borgman, L.E. (1973) "Probabilities for highest wave in hurricane" *J. Waterway Harbors Coastal Engng* **WW2**, 185-207.

Bouws, E., Günther, H., Rosenthal, W. & Vincent, C.L. (1985) "Similarity of the wind wave spectrum in finite depth water: I. Spectral form" *J. Geophys. Res.* **90**(C1), 975-986.

Bretschneider, C.L. (1959) "*Wave variability and wave spectra for wind-generated gravity waves*" *Technical Memorandum #118*, Beach Erosion Board, US Army Corps Engineering.

Bretschneider, C.L. (1972) "A non-dimensional stationary hurricane wave model" *Proceedings of the Fourth Offshore Technical Conference OTC 1517.*

Busch, N.E. & Panofsky, H.A. (1968) "Recent spectra of atmospheric turbulence" *Quart. J. Roy. Meteor. Soc.* **94**, 132-148.

Cardon, V.L. & Pierson, W.J. (1975) "Hindcasting the directional spectra of hurricane generated waves" *Proceedings of the Offshore Technical Conference OTC 2332.*

Case, R.A. (1986) "Atlantic hurricane season of 1985" *Monthly Weather Rev.* **114**, 1390-1405.

Cavanié, A., Arhan, M.K. & Ezraty, E. (1976) "A statistical relationship between individual heights and periods of storm waves" *Proceedings of the Conference on Behaviour Offshore Structure*, vol. 2, pp. 354-360.

Chakrabarti, S.K. & Snider, R.H. (1974) "Wave statistics for March 1968 North Atlantic storm" *J. Geophys. Res.* **79**, 3449-3458.

Charnock, H. (1955) "Wind stress on a water surface" *Quart. J. Roy. Meteor. Soc.* **81**, 639-640.

Chen, Y., Guza, R.T. & Elgar, S. (1997) "Modeling spectra of breaking surface waves in shallow water" *J. Geophys. Res.* **102**(C11), 25035-25046.

Chen, Q., Kirby, J.T., Dalrymple, R.A., Kennedy, B.A. & Chawla, A. (2000) "Boussinesq modeling of wave transformation, breaking and run up. II: 2D" *J. Waterway, Port, Coastal Ocean Engng, ASCE* **126**, 57-62.

Choi, E.C. (1978) "Characteristics of typhoon over the South China Sea" *J. Inst. Aerodyn.* **3**, 353-365.

Collins, J.I. & Viehman, M.J. (1971) "A simplified empirical model for hurricane wind fields" *Proceedings of the Offshore Technical Conference OTC 1346.*

Counihan, J. (1975) "Adiabatic atmospheric boundary layers: a review and analysis of data from the period 1880-1972" *Atmos. Environ.* **9**, 871-905.

Cox, D.T., Kobayashi, N. & Wurjanto, A. (1992) "Irregular wave transformation processes in surf and swash zones" *Proceedings of the 23rd Coastal Engineering Conference*, vol. 1, pp. 156-169.

Davenport, A.G. (1964) "Note on the distribution of the largest value of a random function with application to gust loading" *Proc. Inst. Civil Engng* **28**, 187-198.

Davenport, A.G. (1967) "Gust loading factors" *Proc. Am. Soc. Civil Engng* **ST-3**, 11-33.

Donelan, M.A., Hamilton, J. & Hui, W.H. (1985) "Directional spectra of wind-generated waves" *Philos. Trans. Roy. Soc. Lond. A* **315**, 509-562.

Donoso, M.C.I., LeMehaute, B.L. & Long, R.B. (1987) "Long-term statistics of maximum sea state during hurricanes" *J. Waterway, Port, Coastal Ocean Engng* **113**(6), 636-647.

van Dorn, W.G. (1978) "Breaking invariants in shoaling waves" *J. Geophys. Res.* **83**(C6), 2981-2988.

Doong, D.J., Tsai, L.H., Kao, C.C. & Chuang, L.Z.H. (2001) "A wave spectra study of the typhoon across Taiwan" *Proceedings of the Ocean Wave Measurement and Analysis*, vol. 2, pp. 963-971.

Durst, C.S. (1960) "Wind speeds over short periods of time" *Meteor. Mag.* **89**, 181-187.

Eidsvik, K. (1985) "Large-sample estimates of wind fluctuations over the ocean" *J. Boundary-Layer Meteor.* **32**, 103-132.

Elachi, C.T., Thompson, T.W. & Weissman, D.E. (1977) "Ocean wave pattern under hurricane GLORIA: observations with an airborne synthetic aperture radar" *Science* **198**, 609-610.

Eldeberky, Y. & Battjes, J.A. (1996) "Spectral modeling of wave breaking: application to Boussinesq equations" *J. Geophys. Res.* **101**(C1), 1253-1264.

Finlayson, W.S. (1997) "Spectral growth of hurricane generated seas" *Master's Thesis*, Department of Coastal Engineering, University of Florida, Florida.

Forristall, G.Z. (1978) "On the statistical distribution of wave heights in a storm" *J. Geophys. Res.* **83**, 2353-2358.

Forristall, G.Z. (1988) "Wind spectra and gust factors over water" *Proceedings of the Offshore Technical Conference OTC 5735.*

Forristall, G.Z. & Ward, E.G. (1980) "Directional wave spectra and wave kinematics in hurricanes CARMEN and ELOISE" *Proceedings of the 18th Coastal Engineering Conference*, vol. 1, pp. 567-586.

Forristall, G.Z., Ward, E.G., Cardone, V.J. & Borgman, L.E. (1978) "The directional spectra and kinematics of surface gravity waves in tropical storm DELTA" *J. Phys. Oceanogr.* **8**, 888-909.

Foster, E.R. (1982) "JONSWAP spectral formulation applied to hurricane-generated seas" *University of Florida Report UFL/COEL-82/004.*

Freilich, M.H. & Guza, R.T. (1984) "Nonlinear effects on shoaling surface gravity waves" *Philos. Trans. Roy. Soc. Lond. A* **31**, 1-41.

Garrat, J.R. (1977) "Review of drag coefficient over oceans and continents" *Monthly Weather Rev.* **105**, 915-926.

Gonzalez, F.I., Thompson, T.W., Brown, W.E. & Weissman, D.E. (1982) "SEASAT wind and wave observations of Northeast Pacific hurricane IVA, August 13, 1978" *J. Geophys. Res.* **87**, 3431-3438.

Graham, H.E. & Nunn, D.E. (1959) "Meteorological considerations pertinent to standard project hurricane, Atlantic and Gulf coasts of the United States" *National Hurricane Research Project Report 33*, US Department of Commerce.

Hamm, L., Madsen, P.A. & Peregrine, D.H. (1993) "Wave transformation in the nearshore zone: a review" *Coastal Engng* **21**, 5-39.

Hasselmann, K., *et al* (1973) "Measurements of wind-wave growth and swell decay during the Joint North Sea Wave Project (JONSWAP)" Deutsches Hydrograph Institute.

Holhuijsen, L.H., Booij, N. & Ris, R.C. (1993) "A spectral wave model for the coastal zone" *Proceedings of the Second Ocean Wave Measurement and Analysis*, pp. 631-641.

Holland, G.J. (1980) "An analytical model of the wind and pressure profiles in hurricanes" *Monthly Weather Rev.* **108**, 1212-1218.

Holt, B. & Gonzalez, F.I. (1986) "SIR-B observations of dominant ocean waves near hurricane JOSEPHINE" *J. Geophys. Res.* **91**, 8595-8598.

Huang, N.E., Long, S.R., Tung, C.C. & Yuan, Y. (1983) "A non-Gaussian statistical model for surface elevation of non-linear random wave field" *J. Geophys. Res.* **88**(C12), 7597-7606.

Ishizaki, H. (1983) "Wind profiles, turbulence intensities and gust factors for design in typhoon-prone region" *J. Wind Engng Ind. Aerodyn.* **13**, 55-66.

Iwata, K. & Sawaragi, T. (1982) "Wave deformation in the surf zone" *Memoirs of the Faculty of Engineering, Nagoya Universtiy*, vol. 34, No. 2.

Johnson, A. & Renwick, S.M. (1978) "Buoy observation during hurricane ANITA and BABE" *NOAA Data Buoy Office, Report F-170-2.*

Johnson, A. & Renwick, S.M. (1981) "Buoy observation during hurricane FREDERICK" *NOAA Data Buoy Office, Report F-170-5.*

Johnson, A. & Spear, G.A. (1978) "Buoy observations during hurricane BELLE" *NOAA Data buoy Office, Report F-170-1.*

Kac, M. & Siegert, A.J.F. (1947) "On the theory of noise in radio receivers with square law detectors" *J. Appl. Phys.* **8**, 383-397.

Kaimal, J.C. (1973) "Turbulence spectra, length scales and structural parameters in the stable surface layer" *Boundary-Layer Meteor.* **4**, 289-309.

Kaimal, J.G., Wyngaard, J.G., Izumi, Y. & Coté, O.R. (1972) "Spectral characteristics of surface layer turbulence" *Quart. J. Meteor. Soc.*, 563-589.

Kaminsky, G.M. & Kraus, N.C. (1993) "Evaluation of depth-limited wave breaking criteria" *Proceedings of the Second Ocean Wave Measurement and Analysis*, pp. 180-193.

Karambas, Th. & Koutitas, C. (1992) "A breaking wave propagation model based on the Boussinesq equations" *Coastal Engng* **18**, 1-19.

Kareem, A. (1985) "Structure of wind field over the ocean" *Proceedings of the International Workshop on Offshore Winds and Icing*, pp. 225-235.

Kennedy, A.B., Chen, Q., Kirby, J.T. & Dalrymple, R.A. (2000) "Boussinesq modeling of wave transformation, breaking and run up. I: 1D" *J. Waterway, Port, Coastal Ocean Engng* **126**(1), 39-47.

Kimura, A. & Iwata, Y. (1978) "Wave length, wave velocity and shoaling characteristics of random waves" *Proceedings of the 16th Coastal Engineering Conference*, vol. 1, pp. 320-338.

King, D.B. & Shemdin, O.H. (1978) "Radar observations of hurricane wave direction" *Proceedings of the 16th Coastal Engineering Conference*, vol. 1, pp. 209-226.

Kitaigorodskii, S.A. (1983) "On the theory of the equilibrium range in the spectrum of wind-generated gravity waves" *J. Phys. Oceanogr.* **13**(5), 816-827.

Kitaigorodskii, S.A., Krasitskii, V.P. & Zaslavskii, M.M. (1975) "On Phillips' theory of equilibrium range in the spectra of wind-generated gravity waves" *J. Phys. Oceanogr.* **5**(3), 410-420.

Knowles, C.E. (1982) "On the effects of finite depth on wind-wave spectra: 1. A comparison with deep-water equilibrium range slope and other spectral parameters" *J. Phys. Oceanogr.* **12**, 556-568.

Kobayashi, N., Herrman, M.N., Johnson, B.D. & Orzech, M.D. (1998) "Probability distribution of surface elevation in surf and swash zones" *J. Waterway, Port, Coastal Ocean Engng* **124**(1), 99-107.

Komar, P.D. & Gaughan, M.K. (1972) "Airy wave theory and breaker height prediction" *Proceedings of the 13th Coastal Engineering Conference*, vol. 1, pp. 405-418.

Komen, G.L., Cavaleri, L., Donelan, M., Hasselmann, M., Hasselmann, K. & Janssen, P.A. (1994) "*Dynamics and Modelling of Ocean Waves*" Cambridge University Press, Cambridge.

Kraft, R.H. (1961) "The hurricane's central pressure and highest wind" *Marine Weather Log* **5**, 157.

Krayer, W.R. & Marshall, R.D. (1992) "Gust factors applied to hurricane winds" *Bull. Am. Meteor. Soc.* **73**, 613-617.

Langley, R.S. (1987) "A statistical analysis of non-linear random waves" *Ocean Engng* **14**(5), 389-407.

Leadon, M.E., Nguyen, N.T. & Clark, R.R. (1998) "Hurricane OPAL: beach and dune erosion and structural damage along the panhandle coast of Florida" *Report BCS-98-1*, Bureau of Beach and Coastal Systems, Department of Environmental Protection, Florida State.

Leavitt, E. (1975) "Spectral characteristics of surface layer turbulence over the tropical ocean" *J. Phys. Oceanogr.* **5**, 157-163.

Lee, Y.K. (1980) "Hurricane ELIOSE wave spectra" *Coastal Engng* **4**, 151-156.

Longuet-Higgins, M.S. (1963) "The effect of nonlinearities on statistical distributions in theory of sea waves" *J. Fluid Mech.* **17**, 459-480.

Longuet-Higgins, M.S. (1969) "On wave breaking and the equivalent spectrum of wind-generated waves" *Proc. Roy. Soc. Lond. A* **310**, 151-159.

Longuet-Higins, M.S. (1983) "On the joint distribution of wave periods and amplitudes in a random wave field" *Proc. Roy. Soc. Lond. A* **389**, 241-258.

Lumley, J. & Panofsky, H.A. (1964) "*The Structure of Atmospheric Turbulence*" Interscience Publishers, New York.

McClellan, R.D. (1999) "Method of estimating the nearshore sea severity at the time of hurricane landing as applied to the Florida coast" *Master's Thesis*, Department of Coastal Engineering, University of Florida.

Madsen, P.A. & Schäffer, H.A. (1998) "Higher order Boussinesq-type equations for surface gravity waves: derivation and analysis" *Philos. Trans. Roy. Soc. Lond.* **356**, 1-59.

Madsen, P.A. & Sørensen, O.R. (1992) "A new form of the Boussinesq equation with improved linear dispersion characteristics. Part 2: a slowly-varying bathymetry" *Coastal Engng* **18**, 183-205.

Madsen, P.A. & Sørensen, O.R. (1993) "Boundary waves and triad interactions in shallow water" *J. Ocean Engng* **20**, 359-388.

Madsen, P.A., Murray, R. & Sørensen, O.R. (1991) "A new form of the Boussinesq equation with improved linear dispersion characteristics" *Coastal Engng* **15**, 371-388.

Madsen, P.A., Sørensen, O.R. & Schäffer, H.A. (1994) "Time domain modeling of wave breaking, run-up and surf beats" *Proceedings of the 24th Coastal Engineering Conference*, pp. 399-411.

Madsen, P.A., Sørensen, O.R. & Schäffer, H.A. (1997) "Surf zone dynamics simulated by a Boussinesq-type model. Part 1" *Coastal Engng* **32**, 255-287 (see also Part 2, 289-319).

Miche, R. (1951) "Le pouvoir réfléchissant des ouvrages maritime expóses à l'action de la houle" *Annales ponts et Chaussees* **121**, 285-319.

Michell, J.H. (1893) "The highest waves in water" *Philos. Mag.* **36**, 430-437.

Miller, H.C. & Vincent, C.L. (1990) "FRF spectrum: TMA with Kitaigorodskii's f^{-4} scaling" *J. Waterways, Port, Coastal Ocean Engng* **116**, 57-78.

Miyake, M., Stewart, R.W. & Burling, R.W. (1970) "Spectra and cospectra of turbulence over water" *Quart. J. Roy. Meteor. Soc.* **96**, 138-143.

Mizuguchi, M. (1982) "Individual wave analysis of irregular wave deformation in the nearshore zone" *Proceedings of the 18th Coastal Engineering Conference*, vol. 1, pp. 485-504.

Munk, W.H. (1949) "The solitary wave theory and its application to surf problems" *Ann. NY Acad. Sci.* **51**, 376-462.

NOAA National Data Buoy Center (1986) "Buoy and C-man observations during 1985 Atlantic hurricane season".

Nwogu, O. (1993) "Alternative form of Boussinesq equations for nearshore wave propagation" *J. Waterway, Port, Coastal Ocean Engng* **119**, 618-638.

Ochi, M.K. (1973) "On prediction of extreme values" *J. Ship Res.* **17**(1), 29-37.

Ochi, M.K. (1979) "A series of JONSWAP wave spectra for offshore structure design" *Proceedings of the Behavioral Offshore Structure Conference*, pp. 75-86.

Ochi, M.K. (1993) "On hurricane-generated seas" *Proceedings of the Second Conference on Ocean Wave Measurement and Analysis*, pp. 374-387.

Ochi, M.K. (1998a) "Probability distribution of peaks and troughs of non-Gaussian random processes" *J. Prob. Engng Mech.* **13**(4), 291-298.

Ochi, M.K. (1998b) "Ocean Waves, The Stochastic Approach", Cambridge University Press, Cambridge.

Ochi, M.K. & Ahn, K. (1994a) "Probability distribution applicable to non-Gaussian random process" *J. Prob. Engng Mech.* **9**(4), 255-264.

Ochi, M.K. & Ahn, K. (1994b) "Non-Gaussian probability distribution of coastal waves" *Proceedings of the 24th Coastal Engineering Conference*, vol. 1, pp. 482-496.

Ochi, M.K. & Hubble, E.N. (1976) "On six-parameter wave spectra" *Proceedings of the 15th Coastal Engineering Conference*, vol. 1, pp. 301-328.

Ochi, M.K. & Malakar, S.B. (2001) "Transformation of wave spectra from deep to shallow water" *Proceedings of the Ocean Wave Measurement and Analysis*, vol. 1, pp. 700-713.

Ochi, M.K. & Shin, Y.S. (1988) "Wind turbulent spectra for design of offshore structures" *Offshore Technical Conference OTC 5736.*

Ochi, M.K. & Tsai, C.H. (1983) "Prediction of occurrence of breaking waves in deep water" *J. Phys. Oceanogr.* **13**(11), 2008-2019.

Ochi, M.K. & Wang, W.C. (1984) "Non-Gaussian characteristics of coastal waves" *Proceedings of the 19th Coastal Engineering Conference*, vol. 1, pp. 516-531.

Ochi, M.K., Malakar, S.B. & McClellan, R.D. (2000) "Sea severity at hurricane landing on the Florida coast" *University of Florida Report UFL/COEL-2000/005.*

Olesen, H.R., Larsen, S.E. & Hojstrup, J. (1984) "Modeling velocity spectra in the lower part of the planetary boundary layer" *J. Boundary-Layer Meteor.* **29**, 285-312.

Patterson, M.M. (1974) "Oceanographic data from hurricane CAMILLE" *Proceedings of the Offshore Technical Conference OTC 2109.*

Phillips, O.M. (1958) "The equilibrium range in the spectrum of wind-generated waves" *J. Fluid Mech.* **4**, 785-790.

Phillips, O.M. (1966) "*The Dynamics of the Upper Ocean*", Cambridge University Press, Cambridge.

Pierson, W.J. & Moskowitz, L. (1964) "A proposed spectra form for fully developed wind seas based on the similarity theory of S.A. Kitaigorodskii" *J. Geophys. Res.* **69**(24), 5181-5190.

Pond, S., Phelps, G.T., Paquin, J.E., McBean, G. & Stewart, R.W. (1971) "Measurements of the turbulent fluxes of momentum, moisture and sensible heat over the ocean" *J. Atmosph. Science.* **28**, 901-917.

Powell, M.D. (1980) "Evaluation of diagnostic marine boundary-layer models applied to hurricanes" *Monthly Weather Rev.* **108**, 757-766.

Raubenheimer, B., Guza, R.T. & Elgar, S. (1996) "Wave transformation across the inner surf zone" *J. Geophys. Res.* **101**(C10), 25589-25597.

Rice, S.O. (1945) "Mathematical analysis of random noise" *Bell Syst. Tech. J.* **24**, 46-157.

Robillard, D.J. & Ochi, M.K. (1996) "Transition of stochastic characteristics of waves in the nearshore zone" *Proceedings of the 25th Coastal Engineering Conference*, vol. 1, pp. 878-888.

Ross, D. (1976) "A simplified model for forecasting hurricane generated waves" *Bull. Am. Meteor. Soc.* **57**.

Ross, D. (1979) "*Observing and predicting hurricane wind and wave conditions*" Atlantic Oceanography and Meteorological Laboratory, NOAA, Collected reprints, pp. 309-321.

Schäffer, H.A., Madsen, P.A. & Deigaard, R. (1993) "A Boussinesq model for wave breaking in shallow water" *Coastal Engng* **20**, 185-202.

Schloemer, R.W. (1954) "Analysis and synthesis of hurricane wind patterns over Lake Okechobee" *Florida Hydrometric Report 31*, US Weather Bureau.

Schmitt, K., Friehe, C.A. & Gibson, C.H. (1979) "Structure of marine surface layer turbulence" *J. Atmos. Sci.* **36**, 602-618.

Schroeder, J.L., Smith, D.A. & Peterson, R.E. (1998) "Variation of turbulent intensities and integral scales during the passage of a hurricane" *J. Wind Engng Ind. Aerodyn.* **77/78**, 65-72.

Schroeder, J.L. & Smith, D.A. (1999) "Hurricane BONNIE wind flow characteristics as determined from WEMITE" *Proceedings of the 20th Conference on Wind Engineering*, vol. 1, pp. 329-335.

Schwerdt, R.W., Ho, F.P. & Watkins, R.A. (1979) "Meteorological criteria for standard project hurricane and probable maximum hurricane wind field, Gulf and East coast of the United States" *NOAA Technical Report NWS-23.*

SethuRamen, S. & Raynor, G.S. (1975) "Surface drag coefficient dependence on the aerodynamic roughness of the sea" *J. Geophys. Res.* **80**(36), 4983-4987.

Shears, M. & Bell, M.H. (1979) "Dynamic effects of wind on offshore towers" *Proceedings of the First Conference on Environment Force on Engineering Structures*, pp. 61-74.

Shemdin, O.H. (1980) "Prediction of dominant wave properties ahead of hurricanes" *Proceedings of the 17th Coastal Engineering Conference*, pp. 600-609.

Shiraishi, N. (1960) "Aerodynamic response of bridges to natural winds" *Proceedings of the Second US–Japan Seminar on Wind Effects on Structures.*

Simiu, E. & Leigh, S.D. (1983) "Turbulent wind effects on tension-leg platform surge" *National Bureau Standards, Building Series 151.*

Sneider, R.H. & Chakrabarti, S.K. (1973) "High wave conditions observed over the North Atlantic in March 1968" *J. Geophys. Res.* **78**, 8793-8807.

Sobey, R.J., Chandler, B.D. & Harper, B.A. (1990) "Extreme waves and wave counts in a hurricane" *Proceedings of the 22nd Coastal Engineering Conference*, vol. 1, pp. 359-370.

SWAMP Group (1985) "*Ocean Wave Modeling*" Plenum Press, New York.

Takeda, A. (1981) "Effects of water wave motion on spectral characteristics of wind fluctuation in the marine atmospheric surface layer" *J. Meteor. Soc. Jpn* **59**(4), 487-508.

Tayfun, M.A. (1980) "Narrow-banded nonlinear sea waves" *J. Geophys. Res.* **85**(C3), 1548-1552.

Thornton, E.B. (1977) "Rederivation of the saturation range in the frequency spectrum of wind generated gravity waves" *J. Phys. Oceanogr.* **7**, 137-140.

Thornton, E.B. & Guza, R.T. (1983) "Transformation of wave height distribution" *J. Geophys. Res.* **88**(C10), 5925-5938.

Tieleman, H.W. (1995) "Universality of velocity spectra" *J. Engng Wind Ind. Aerodyn.* **56**, 55-69.

Walsh, E.J., Wright, C.W., Vandemark, D., Krabill, W.B., Garcia, A.W., Houston, S.H., Murillo, S.T., Powell, M.D., Black, P.G. & Mark, E.D. (2002) "Hurricane directional wave spectrum spatial variation at landfall" *J. Phys. Oceanogr.* **32**, 1667-1684.

Wang, G.C.Y. (1978) "Sea-level pressure profile and gusts within a typhoon circulation" *Monthly Weather Rev.* **106**, 954-960.

WAMDI Group (1988) "The WAM Model – a third generation ocean wave prediction model" *J. Phys. Oceanogr.* **18**, 1775-1810.

Ward, E.C., Borgman, L.E. & Cardon, V.J. (1979) "Statistics of hurricane waves in the Gulf of Mexico" *J. Petrol. Tech.*, 632-642.

Wei, G., Kirby, J.T., Grilli, S.T. & Subramanya, R. (1995) "A fully nonlinear Boussinesq model for surface waves. I: highly nonlinear, unsteady waves" *J. Fluid Mech.* **294**, 71-92.

Wei, G., Kirby, J.T. & Sinha, A. (1999) "Generation of waves in Boussinesq models using a source function model" *Coastal Engng* **36**, 271-299.

Weishar, L.L. & Byrne, R.L. (1978) "Field study of breaking wave characteristics" *Proceedings of the 16th Coastal Engineering Conference*, pp. 487-506.

Whalen, J.E. & Ochi, M.K. (1978) "Variability of wave spectral shapes associated with hurricanes" *Offshore Technical Conference OTC 3228.*

Withee, G.W. & Johnson, A. (1975) "*Buoy observation during hurricane ELOISE*", NOAA Data Buoy Office, NSTL Station.

Wright, C.W., Walsh, E.J., Vandemark, D., Krabill, W.B., Garcia, A.W., Houston, S.H., Powell, M.D., Black, P.G. & Marks, F.D. (2001) "Hurricane directional wave spectrum variation in the open ocean" *J. Phys. Oceanogr.* **31**, 2472-2488.

Wu, J. (1980) "Wind-stress coefficients over sea surface near neutral condition" *J. Phys. Oceanogr.* **10**(5), 727-740.

Wu, J. (1982) "Wind-stress coefficients over sea surface from breeze to hurricane" *J. Geophys. Res.* **87**(C12), 9704-9706.

Young, I.R. (1988) "A shallow water spectral model" *J. Geophys. Res.* **93**, 5113-5129.

Young, I.R. (1988) "Parametric hurricane wave prediction model" *J. Waterway, Port, Coastal Ocean Engng* **114**(5), 637-652.

Young, I.R. (1998) "Observations of the spectra of hurricane generated waves" *Ocean Engng* **25**, 261-276.

Young, I.R. & Burchell, G.P. (1996) "Hurricane generated waves as observed by satellite" *Ocean Engng* **23**, 761-776.

Zelt, J.A. (1991) "The run-up of non-breaking and breaking solitary waves" *Coastal Engng* **15**, 205-246.

Index